William Peck

The Constellations and How to Find Them

13 maps, showing the position of the constellations in the sky during each

month of any year. A popular and simple guide to a knowledge of the

starry heavens. Third Edition

William Peck

The Constellations and How to Find Them
13 maps, showing the position of the constellations in the sky during each month of any year. A popular and simple guide to a knowledge of the starry heavens. Third Edition

ISBN/EAN: 9783337248444

Printed in Europe, USA, Canada, Australia, Japan

Cover: Foto ©berggeist007 / pixelio.de

More available books at **www.hansebooks.com**

THE CONSTELLATIONS SURROUNDING THE NORTH POLE.

ALWAYS VISIBLE IN THE NORTHERN HEMISPHERE.

THE above Map is a representation of the Starry Heavens near the North Pole. These Stars are always to be seen on every clear night, as they never sink below our Northern Horizon, but appear to revolve round the Pole in the same time that the Earth takes to rotate on its axis, and in an opposite direction from the hands of a watch.

For a description of this Map, and the manner of using it see page 11.

THE

CONSTELLATIONS

AND

HOW TO FIND THEM

13 MAPS, SHOWING THE POSITION OF THE CONSTELLATIONS IN THE SKY
DURING EACH MONTH OF ANY YEAR

A POPULAR AND SIMPLE GUIDE TO A KNOWLEDGE
OF THE STARRY HEAVENS

WITH INTRODUCTION, GENERAL EXPLANATIONS, AND A SEPARATE DESCRIPTION OF EACH MAP

BY WILLIAM PECK, F.R.A.S.

THIRD EDITION.

Silver, Burdett & Co.,
PUBLISHERS,
6 HANCOCK AVE., BOSTON, MASS.

CONTENTS.

THE CONSTELLATIONS AND HOW TO FIND THEM.

INTRODUCTION.

FROM the very earliest ages the stars have been watched with interest and admiration, and their movements traced out and applied to various useful purposes. Their influences, too, on the fortunes and destinies of man, were made the subject of ignorant and superstitious enquiry, as it was believed in those early times that the "stars in their courses" ruled the fate of men and nations. Nor can it be wondered at, that long before the motions of the heavenly bodies were accurately known, men would look up to the starry heavens in wonder and reverence, watching with superstitious awe those apparently innumerable orbs. They would naturally believe that by a knowledge of the stars and their movements, they would be able to foretell future events with great exactness, and to think that the ever-varying aspects of the heavens, in their regular progression, and solemn and stedfast silence, would, if studied, reveal to them the secret of their future destinies. But even at

Fig. 1. Southern sky 4000 years ago. the present time, when science and religion have so enlightened the world, there are those who, though they do not believe in the influences of the stars, or astrology, believe in planetology, or in the power that planets and comets are supposed to have over famines, pestilences, droughts, earthquakes, and such like, which is just as ridiculous as the ancient system of astrology. It is, however, not my purpose at present to discuss the difference between ancient and modern astrology, but rather to shew how a knowledge of the principal stars and constellations may be easily attained, and to point out anything that may be of interest in connection with them.

To any one who has an interest in the study of the starry heavens, there is no occupation more agreeable than to observe the sky on a clear night, and watch the varied positions of the constellations from season to season; noting as the months advance familiar stars disappearing in the west as new groups appear in the east, till after the lapse of a year the heavens will again represent the same appearances.

It was evidently for the purpose of identifying the stars and finding out more about them that the first watchers of the sky divided the heavens into groups, or constellations, naturally naming each group after some object to which they fancied it had a resemblance. As these first observers of the heavens were chiefly shepherds or herdsmen, we can readily conceive how the oldest constellations are generally called after objects and animals with which a herdsman would be familiar in those early times. They would picture to themselves in the different star-groups the objects with which they would be best acquainted; and thus it is that we see scattered all over the heavens groups of stars representing bears, lions, sheep, and oxen; and even the herdsman and huntsman himself.

It was undoubtedly from the Chaldæans and Egyptians that we derived this system of naming and recognising the constellations, although the Arabians, Persians, Greeks, and others have added many constellations of their own; and even in modern times a great many new names have, from time to time, been given to small groups of stars, which have not always been accepted by uranographers. The idea of the constellation figures is evidently very old, for there are few ancient authors in which some of them, at least are not to be found. Bootes and the Bear are mentioned both by Homer and Hesiod; and Job—who is supposed to have been an Arabian chief prior to the time of Moses—speaks about Arcturus, Orion, and the Pleiades, so that a great many of the constellations with which we are familiar at the present day were known to the people who lived in those early ages. It is probably more than 4,000 years ago since the oldest star groups were first named, and by a people—as it is thought by some astronomers—who lived in a country at no great distance from Mount Ararat. At that time—when the present names of the constellations were first invented—the heavens did not present the same appearance as they do at the present day; for we know that the earth, besides rotating on its axis and revolving round the sun, reels like a mighty gyroscope, but with so slow a motion that it takes nearly 25,900 years to make one complete revolution of its axis round an imaginary line perpendicular to the plane in which the earth moves. Still further, as this axis of the earth moves in its circuit round this perpendicular line, it points successively to different parts of the heavens, and as this point in the heavens to which the axis is directed (called the celestial pole) will not have any diurnal motion, all the stars will appear to revolve round it, or round the star that may be nearest to it; from which circumstance it will be called the pole star. Thus 4000 years ago, the earth's axis pointed in a direction different from what it does at present, and therefore the same constellations at that time would appear in a different part of the heavens. The axis or *pole* was then pointing to a star in the constellation of Draco, called *Thuban*, which was the ' Pole Star' at the time when the star-groups are supposed to have been first named. It was also the Pole Star to the Egyptians at the time when the great Pyramid of Cheops was built, and no doubt it would be of great use in building that structure; which is found to be very accurately placed with regard to the four cardinal points.

The Pole star at that time would shine down the long slanted tunnel in the side of the Pyramid, and would probably be seen shining in the day time as well as at night.*

Fig. 2.—The North Polar heavens 4,000 years ago.

In Plate A there will be seen a dotted circle, whose circumference crosses very near *Polaris*; this circle shows the path that the North Pole of the earth traces out in the heavens, travelling in the direction of the arrows, and making one complete revolution in 25,900 years. The circle is divided into equal spaces, each of which represents 1,000 years, so that it can be easily found where the Pole will be in the future, or where it was in the past. It will be seen that this path of the Pole passes very near Thuban, shewing that at one time the axis of the earth would point to that star, which it did, as before mentioned, about 4,000 years ago; and it will be further seen that the Pole has not yet reached its nearest to alpha of the Little Bear, which it will do 300 years hence; and also that in about 13,000 years (half a revolution of the Pole) from the present, the bright star Vega will occupy the same position, with regard to the Pole, as Polaris now does.

Fig. 2. is a representation of the North Polar heavens when Alpha Draconis was nearest the Pole. It will be seen that at that time the constellation of Draco, the Dragon, occupied the furthest North part of the celestial sphere;— and thus would hold a prominent position in the heavens—the star that was then brightest in this constellation (Thuban) would be the Pole Star of the period, 4,000 years ago.

In passing, as it may be of some interest to the reader, we may mention that there are those who believe that when ancient writers speak about the "Old Dragon" being cast out of heaven and "his tail drawing the third part of the stars of heaven," that reference was made, perhaps unconsciously, to an old tradition of Draco falling away from his prominent position among the constellations, which, by the gyratory motion of the earth, he has really done, taking the surrounding stars with him. The above persons also see in a whole series of constellations near each other the story of the deluge; in the ship Argo they see the ark; in Aquarius, the

---

* It is from the tunnel in the great Pyramid that the date of its building is known, far beyond doubt, the tunnel was made so as to point to the Pole Star of the period, and by calculating back (within reasonable limits) it is found that there is only one bright star that could have been in the position to shine down the long tunnel; that star is Alpha of the Dragon, *Thuban*, which would be in the above position in the year 2170 B.C., the year (rather century) that the Pyramid was built.

Water-Bearer (which represents a man pouring water out of a vessel); and in the river Eridanus, the flood is represented, with Pisces, the fishes, and Cetus, the whale, swimming in the "deep waters." They also see in Corvus and in Columba, the raven and the dove that Noah sent out of the ark. Again in Centaurus—which originally represented a man offering a sacrifice—Noah is perceived offering a sacrifice after leaving the ark; and in the bow of Sagittarius, the bow of promise, set in the clouds above the altar, which is represented in the constellation of Ara.

The above plan of the ancients in connecting with the different star-groups, names suggestive of the great events of the early history of the world, seems easily accounted for, when we remember the fact that the constellations were likely named (perhaps a few centuries) after the deluge, when men's minds were still, from traditionary tales, full of the terrible visitations which the earth had witnessed, and so would easily suppose that they saw in the heavens, outlined by stars, the whole narrative of Noah's flood.

Those who thus found a picture of the deluge in these constellations, also found in Hercules, defeated by the serpent, the first Adam; and in Ophiuchus, the serpent-bearer, the type of the second Adam triumphant over the serpent; and in Orion, the noblest of all the constellations, Nimrod, "the mighty hunter," with his two dogs, Canis Major and Canis Minor, and the animals he hunted, such as Lepus, the hare.*

Nearly all the constellations, as they are at present depicted, have no resemblance whatever to the objects which gave rise to their names, but this can, in many instances, be easily accounted for. The motion of the earth's axis round the Pole of the ecliptic, as already described, would greatly alter the positions of the stars, with regard to the diurnal motion of the heavens, which would much affect the general appearance of the constellations as compared with their appearance at present. In the case of Argo, for instance, which at present is never seen on a horizontal keel, as one would naturally suppose a ship should always be seen, but, about 4,000 years ago, in the latitude of Chaldæa, or Egypt, it would be seen just above the southern horizon, sailing horizontally; so that this constellation is, owing to the gyratory motion of the earth, never seen by us in the same manner as it was seen at the time when it was first named. Again, as the representations of the constellations have been handed down to us, they have, in a great many instances, been robbed of some of their principal stars, which have gone to form new constellations, and in other cases, stars have been added to constellations, which belonged to other asterisms; so that now we do not, for certain, know what actually were all the stars belonging to the old constellations.

The difference between the ancient and modern appearance of the constellations may still further be accounted for on the supposition that the astronomers (rather astrologers) who divided the heavens into groups, would not require that the stars of a constellation should belong exclusively to it, but would look upon a great many of the stars as common to several constellations. When, however, the exact position of a star in the heavens was required (which may have been the case as far back as the time of Hipparchus), each constellation would require to have boundaries assigned in order to know what stars really belonged to it; thus, any star that was before common to many groups, would no longer require to be so, but would, henceforth, be exclusively fixed to one constellation. This will account in some manner for the lost stars of some of the constellations, without which the group would never have derived its present name. There are instances of this in various parts of the heavens—in Hercules the star Iota would, in all probability, belong to Draco, when that constellation was named. The same occurs in Coma Bernices (a small group of stars near the tail of Leo which was named by Tycho Brahe in 1604) which at one time would likely form the tuft of hair in the tail of the Lion, and in the modern constellation of Crux (which is only seen in the southern hemisphere), it is thought by Mr. Proctor that it originally belonged to Centaurus.

The above suppositions may account for the fact that so many constellations now have not the least resemblance to the objects that they are supposed to represent, and may likewise explain how the original constellations would, when first imagined, have some likeness to the objects from which they were named. Ursa Major, the Great Bear, for instance, must originally have had a striking resemblance to a bear, for it was recognised quite independently by many nations besides the ancient Greeks. Draco still has some likeness to a dragon (which will be seen by glancing at Fig. 2); and there are other constellations besides these, which have at present a resemblance to the object from which they derived their names—such as the Northern Crown, Dolphinus, Scorpio, &c.

* For a full account of the above see Mr. R. A. Proctor's interesting book "The Myths and Marvels of Astronomy."

It was probably from the Chaldæans that the Egyptians derived their knowledge of the constellations, for the architects of the Great Pyramid are thought to have been of that race, and they have shown us by the records they have left in that huge structure that they were as far advanced in astronomy and mathematics as could be expected in those very early ages. Indeed, it is highly probable that the Chaldæans, at the time the Pyramid was built, possessed a far more accurate knowledge of astronomy than the Greeks did 2,000 years afterwards.

The Egyptians would likely communicate the knowledge of the old constellations, and astronomy, as it was then known, to the surrounding nations—to the Arabians, Persians, and to the ancient Greeks, &c.—each nation adding some constellations of its own. The Arabians gave individual names to the brightest stars; generally naming the star from the position it occupied in the constellation; as, for instance, Betelgeux from Ibt-al-Jauza, the giant's shoulder—that star being on the shoulder of the "Giant Orion."

The ancients used the stars for many useful purposes, in fact the stars were to the people, who lived two or three thousand years ago, what our almanacs are to us at the present day, for by means of the stars they knew when it was a certain time of the year, whether seed time, or harvest, or the beginning of seasons, &c., all of which would be valuable for many purposes besides agricultural. They watched when certain conspicuous stars rose, or set, immediately before, or after, the sun (called the heliacal rising of a star), which, only taking place once a year for each star, would let them know how far the seasons had advanced. The people in Egypt, for instance, knew when the Nile would likely overflow its banks, by the star Sirius rising heliacally and becoming visible in the early morning. By the stars the Phœnicians (who are supposed to have been the originators of nautical astronomy) knew in what direction to steer their ships at night by watching the Little Bear, which they knew to be always near the North—as Aratus, referring to this constellation, informs us—

"Observing this, Phœnicians plough the main."

The use of the Pole Star (a bright star in the tip of the Little Bear's tail) in navigation is said to have been introduced into Greece by Thales, who derived his knowledge from the Phœnicians; it has ever since been used by sailors for finding their position on the trackless ocean.

In modern astronomy the old constellation names do not hold so important a position as they once did. Still they form a convenient method for distinguishing the stars, though to the practical astronomer the number of a star in some important catalogue (such as "Struve, 3521") is all that is required to identify it. However, to those who only want to be able to distinguish the stars, the old constellations and the individual names of stars, have a far greater interest, and are thus more easily remembered than a star known only by number; and it is highly probable that the constellation figures, which have for so many centuries been fixed upon men's minds, will not be dismissed even from exact astronomy for many years to come; for as the fancied figures in the different star-groups had once such a hold upon mankind, and are so often mentioned by ancient poets, so they will not be altogether forgotten, even after astronomy has adopted some more improved system of distinguishing the stars than at present exists.

In the light of the wonderful discoveries that have been made in the siderial heavens since Galileo first turned the telescope to the stars, each star that is seen on a clear night is now looked upon as a sun, and so will be the centre round which other planets revolve; just as our sun, which is also a star, and far from being the most important among its companions,* is the centre to its family of worlds circling round it—of which the earth is one of the least—so that

"The radiant orbs
That more than deck, that animate the sky,
Are life-infusing suns of other worlds."

But what may at first seem surprising is the fact that, amidst the all-seeming stillness of the star depths, there is yet going on a continual movement and uproar. Each sun is moving with a fearful velocity, and carrying its family of planets along with it. Some are travelling on their stately course at such a speed that every hour they pass over a space equal to *three thousand miles*. Our sun is also travelling, and with a great velocity,† so that in fact

---

* There is every reason to believe that our Sun belongs to an inferior order of suns, and that Sirius belongs to a higher order, and at least exceeds our sun *one thousand* times in volume.

† According to Sir William Herschel our sun with its numerous planets is travelling towards the star Lambda in the constellation of Hercules.

"suns are revolving round suns, and systems round systems." But further, each orb is aglow with fiery energy; it is a storehouse from which bounteous supplies of light and life are continually being given out to the numerous worlds that are kept circling round it by the attraction of its mighty influence. We know in the case of our sun that storms are raging on its surface, in which great masses of intensely heated vapour, thousands of miles in breadth, are rushing onward with dreadful force and at a great velocity—a velocity that is hundreds of times greater than the speed of a cannon ball. We also know of the great masses of glowing matter that are occasionally flung from the sun to a height nearly as great as is half the distance from the earth to the moon. And as these motions and uproars cannot take place without a great noise, there must be incessantly produced on the sun a noise compared to which the loudest crash of thunder, or the roar of the greatest piece of artillery, will be as absolute silence. This, then, being the *seeming* stillness of our sun, what must be the uproar on a sun exceeding ours one thousand times in volume? It will likely be proportionally as great; so that in the awful stillness of the star depths there is going on continually a fearful uproar and tumult, compared to which the greatest noise that we can realise sinks into complete nothingness.

Then in what infinite numbers are the stars scattered throughout the universe and these disturbances going on in each! Each one is, without doubt, working out the purpose of its great Creator, and until that purpose shall be fulfilled it will, like some mighty engine, work unceasingly, giving out life, light, and heat to the many worlds that circle round it. Each is the source from which countless forms of life derive their existence, and as we know that there is, at least, one planet—probably several—in our solar system that is inhabited with intelligent beings, we are naturally led to the conclusion—it would be unreasonable to suppose otherwise—that those planets circling round other suns—many of which far surpass our own in splendour—are the abode of other intelligences. Again, at what enormous distances are the stars situated from each other; the nearest to us, as at present known, being no less than 20 *billions* of miles away, a distance that is so great that it cannot be travelled by light (which moves at the velocity of 190 *thousand* miles each second) in less than three and a half years! But at what a vast distance must those stars be that even light itself requires not only 10, 20, or a 100 years, but 10, 20, or 100 *thousand* years to complete the journey to our earth! We do not see the stars, therefore, as they are now, for of their present existence we have not the slightest knowledge, but as they were 10, 20, or 100 thousand years ago. They may have ceased to exist for many years, because we will not know till the ray of light bearing the information reaches us; it may be on the journey to tell us of a conflagration that has taken place on a star 20 years ago, or that a system of worlds has been destroyed. Seeing, therefore, that light takes such an interval of time to journey from one star to another, we may perceive, though dimly, how vast the dimensions of this visible universe must be.

And this is only the known universe! How great is the unknown? For the many millions of stars which are revealed to us by means of the most powerful telescope, are only as "a drop in the bucket" to the infinite number which exists throughout the boundless universe! In fact, all the stars which are seen in the heavens, together with our Sun, form a part only of *one* vast and complicated aggregation of orbs, of which there are, in all probability, an endless number scattered throughout the depths of infinite space. What, then, must we think when we thus find, not only myriads of stars, but myriads of *star-clusters*, each containing millions of millions of suns—suns, which, perhaps, are far more resplendent, and which rule over hundreds of worlds more spacious than our own.

The mind cannot realise the meaning and infinite significance of such a wondrous scene. And, further, when we reflect that, very probably, each of these worlds, revolving round other suns, will, at some period or another of their existence, be the abode of intelligent life, a deeper meaning is given to the words of the inspired Psalmist,—"When I consider Thy heavens, the work of Thy fingers, the moon and the stars, which Thou hast ordained; what is man, that Thou art mindful of him? and the son of man, that Thou visitest him?"

GENERAL EXPLANATIONS.

THE whole heavens appear to make one complete revolution in 23 hours and 56* minutes from East to West, which apparent motion is caused by the rotation of the earth on its axis in an opposite direction, giving rise to the phenomenon of the rising and setting of the sun and stars. But, besides rotating on its axis, the earth also revolves in an orbit round the sun once in a year, in the same direction that it rotates (*i.e.*, from West to East); from which circumstance the sun will *appear* to move in the same direction among the constellations, making a complete circuit of the stellar heavens in the course of a year.† Now, as it is from the sun that we derive our time, it being always apparent noon when the sun and that part of the heavens that he is in is due South, the stars will appear to move forward to meet the sun; and as the sun is South at 12 o'clock noon, the stars will come earlier to the South every day. As the earth advances in its orbit about one degree (equal to 4 minutes of time) each day, the star sphere will present the same appearances 4 minutes sooner every day; which, in the course of a month, will make a difference of nearly 2 hours earlier, so that by the apparent motion of the heavens towards the West in the course of a year—caused by the real motion of the earth round the sun—the stars will occupy different positions in the sky, according to the different seasons of the year.

For the purpose of following the stars and constellations from month to month throughout the year, I have constructed this series of Maps to show the aspects of the sky at stated times;‡ and they are so arranged that they are ACCURATE FOR EVERY YEAR. Only the principal stars of the most important and conspicuous constellations have been inserted, as to a beginner it would be very confusing if the maps had been crowded with small and faint stars. Lines have been drawn through the stars showing the principal features of each constellation, which, it is hoped, will be found to be of far more assistance in identifying a group of stars in the heavens, than the insertion of the constellation figures would have been, especially as they are at present delineated on some celestial globes and star atlases.

Each of the Maps from I. to XII. represents the position of the constellations that are visible at the day and hour stated at the bottom of each. The circumference of each circle represents the horizon with the eight principal points of the compass inserted, and so as to make the horizon appear more natural, there is roughly represented the objects that are generally seen on the horizon, such as hills, &c. The small cross, seen in the centre of each Map, is that part of the sky which is directly overhead, or zenith as it is called. The Ecliptic and the Equinoctial have been inserted in each, it being thought that these lines would add greatly to the completeness of the Maps without tending to confuse, as the former is the apparent yearly path of the sun, and near which the larger planets are always to be found, and the latter the circle that divides the heavens into two equal parts—the Northern and Southern Hemispheres.

The moon also makes a circuit of the star sphere, but in a much shorter time than the sun; for, whereas the sun takes one year to move once round the heavens, the moon only takes one month (or more accurately 27 days, 7 hours, 43 minutes) to complete the same revolution, so that the moon will travel more than 13 times round the heavens in the same time that the sun takes to move round once. The moon passes through the same constellations as the sun, but being sometimes as far as 10 times her own diameter to the North or South of the sun's path, she will not, therefore, move in it. She crosses it, however, twice every month at two points, called the nodes of her orbit; these points are not always on the same part of the Ecliptic, but make a complete circuit of it in nearly 19 years.

To attain a knowledge of the principal stars and constellations by aid of these Maps is exceedingly simple. All that the observer has to do is to take the Map (see Maps I. to XII.) that represents the constellations for the month wanted, and the hours of observation will be seen at the bottom, though half an hour or so on either side of the time stated will do well enough for the purpose of identifying the stars; only, be it remembered, that if later than the hour mentioned, the stars in the Eastern quarter of the sky will be a little higher, and those in the Western heavens a little lower, than is represented on the Map, and, of course, if the time is earlier than stated, then it will be the opposite of this.

Take then the Map, and face one of the cardinal points (say the North), then turn the map round so that the North part of the horizon is lowest, and the cross or zenith above; you will thus have an exact representation of the Northern heavens. Turning now directly to the South, and keeping the Southern horizon underneath, and the cross vertically above, you will see all the stars that are in that part of the sky, and so on for any other part of the heavens. Thus the principal constellations that are visible at any particular time may be easily found, and their varied positions followed throughout the year. Facing each Map will be seen a description of the constellations that are represented.

* More accurately 23 hours, 56 minutes, 4 seconds.
† Those Constellations among which the Sun moves are called the Zodiac, or "yoke of the sky," they are twelve in number, viz., Aries, Taurus, Gemini, Cancer, Leo, Virgo, Libra, Scorpio, Sagittarius, Capricornus, Aquarius, and Pisces.
‡ These Maps have been constructed on the Globular Projection, as it gives the least distortion for a whole hemisphere.

THE CONSTELLATIONS SURROUNDING THE NORTH POLE.

Plate A shows those stars that never sink below the horizon in Britain, and so they will always be visible on every clear night; but, of course, the same constellations will not always be seen in the same position in the (Northern) sky, for as the months advance the stars will appear to move slowly round a bright star in Ursa Minor, the Little Bear (named Polaris, the 'Pole Star,' which will be seen as the centre of the Plate, or Map), in a direction contrary to the hands of a watch.

To find when this Map coincides with the Northern heavens, turn it round till the day of the month—seen at the circumference of the circle—is at the bottom and Polaris vertically above; then this will be the exact appearance of the Northern sky at 12 o'clock midnight on that day. If this Map is wanted to represent the heavens at an earlier hour, it will then require to be turned to the left, just as many hours from midnight as wanted (each of the short lines at the inside circle being equal to one hour), and if later, the Map will require to be turned the number of hours necessary in an opposite direction. For example :—Suppose that it is desired to set the Map so as to find what constellations are in the various parts of the Northern sky on December 20, at 10 o'clock in the evening. First, find the day of the month in the circumference of the Map, then turn it round so that this date will be at the bottom; but as this will only represent the heavens at midnight on that day, the Map will require to be turned to the left two hours, and this will show the aspect of the sky at that time. Hanging down below the Pole Star will be seen the Little Bear, Ursa Minor; Cassiopeia will be nearly overhead with Perseus; below Ursa Minor the long straggling constellation of Draco will be found, and in the North-West the star Vega, in Lyra, the harp, will be sparkling brightly. Thus by means of the scale of months the Map can be readily turned to the right position.

THE MAGNITUDES OF THE STARS, Etc.

On looking at the heavens it will be seen that the stars do not all shine with the same brilliancy; for some shine with more brilliancy than others. Astronomers have divided all the lucid stars (i.e., all the stars that are visible to the naked eye) into divisions, each division being called a magnitude. Those stars that appear to be the brightest are called stars of the first magnitude; next to these come stars of the second magnitude, and so on till the sixth magnitude stars are reached, which are the smallest stars visible to the naked eye. As the division of stars into classes was made long before the invention of the telescope, the stars that are fainter than the sixth magnitude will only be seen with the assistance of that instrument, and are therefore called 'telescopic stars.' They are classed in magnitudes varying from the 7th to the 18th or 19th; these last magnitudes, of course, will only be visible by means of the most powerful telescopes now in use, nor is there any reason why a limit should be assigned to this progressive diminution, for the past has shown that with every improvement of the telescope fainter stars have been brought to light, and if so in the past, why not the same in the future?

The following table shows the relative amount of light that reaches the earth from stars of the first six magnitudes, the first magnitude star being equal to 100.

First	Magnitude	equal to	100	Fourth	Magnitude	equal to	6
Second	,,	,,	25	Fifth	,,	,,	2
Third	,,	,,	12	Sixth	,,	,,	1

Figs. 3 and 4 have been constructed so as to show all the first six magnitude stars, or all those stars that are visible to the naked eye in the Northern and Southern Hemispheres respectively. The number of stars that are thus seen in the whole heavens is nearly 6,000—about 2,500 in the Northern, and nearly 3,300 in the Southern Hemisphere.

The ancients gave individual names to the most important stars, which was the only method of recognising them till the year 1603, when John Bayer, of Augsburg, introduced a new system of distinguishing the stars, a system that has ever since been employed by astronomers. In this system the names of the old constellations are still kept, but each star is designated by a letter of the Greek alphabet, calling the brightest star in a constellation after the first Greek letter, the second brightest star by the second letter, thus :—α (Alpha) Lyræ is the brightest star in the constellation of Lyra, β (Beta) the second brightest, and so on till all the Greek letters are exhausted. When the number of stars in a constellation exceeds the number of letters in the Greek alphabet, then small Roman letters are used in continuation, thus :—after ω (Omega) the last Greek letter, a, b, c, d, &c. are used, and if more stars still remain, then numerals are resorted to. On page 16 there is given a table showing the Arabic names of the principle stars visible in Britain, with the Greek letters of each constellation.

Fig. 3.
The Stars visible to the naked eye in the Northern Hemisphere.

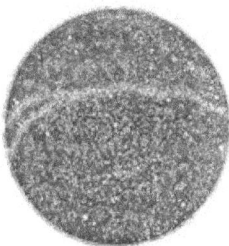

Fig. 4.
The Stars visible to the naked eye in the Southern Hemisphere.

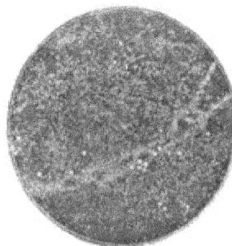

THE PLANETS.

Sometimes it may happen that a very bright star will be seen shining in the heavens, and on looking at the Maps to see to what constellation it belongs, the observer will at first be disappointed to find that it has not been inserted, and so will naturally suppose that the Maps are imperfect in that respect. On a closer inspection of the stranger, however, to see what position it would occupy in the Maps (which is easily done by noting if it forms a triangle or line, &c., with known stars), it will be found that it is not far away from the Ecliptic, or sun's path, and in this case it is sure to be one of the planets (*i.e.*, wandering stars). If the unknown star shines with a clear steady light, and at the same time is not far away from the sun (*i.e.*, rising shortly before or setting soon after the sun), it will probably be the planet Venus.

If it is shining with a bright clear light and is situated at a good distance from the sun—seen for instance near the South at midnight—it will without doubt be the giant planet Jupiter, the largest planet in the solar system. When the stranger appears as a very red star it will doubtless be the planet Mars, and if seen shining with a dull greenish colour, it will be the ringed planet Saturn.

As each of the planets move among the stars, some more rapidly than others, it will at once be understood how it is that they have not been inserted in the accompanying Maps ; for if they had been placed in their proper positions for this year in the monthly Maps, they would by the next year have moved from these positions, and, consequently, the Maps would *not* have been ACCURATE FOR EVERY YEAR, which without the planets they *really* are.

PLATE 8.

THE CONSTELLATIONS SURROUNDING THE SOUTH POLE.

NEVER VISIBLE IN THE NORTHERN HEMISPHERE.

THE above Map is a representation of the Starry Heavens near the South Pole. These Stars are not seen in Britain, as they never appear above our Southern Horizon.

For a description of this part of the Heavens see page 13.

THE CONSTELLATIONS SURROUNDING THE SOUTH POLE.

Plate B is a representation of that part of the Southern heavens which, being always below the horizon in Britain, is never seen. This Plate has been inserted so that those who have not seen the Southern skies may have an intelligent idea of the stars and constellations with which our friends are familiar in the Southern Hemisphere. They have not a constellation like our Great Bear always pointing to a bright Pole star, but they have a "Southern Cross," which rivals in grandeur any group of stars with which we are familiar. This conspicuous constellation will be seen near the top of the Map towards the left, just on the edge of the Milky Way ; the two stars a and γ are the Southern 'Pointers,' for a line from γ through a will point out the South "Pole Star," σ (Sigma) of Octans, which star being very faint will only be found with difficulty. Crux is a modern constellation, it being named by Royer in 1679 ; in fact, the greater part of the Southern constellations have been named in modern times and within the last 300 years. Bayer named 12, Hevelius 2, Royer 5, Halley 1 (Charles' Oak), and La Caille 14. All these, however, have not been accepted by astronomers, only the most important and larger groups being retained. But how is it that this particular part of the heavens has not been divided into constellations before? Why are all these constellations modern? We see in other parts of the heavens groups that were named thousands of years ago, and that some of the constellations with which we are familiar were known to the Chaldæans long before Greece and Rome were in existence. When these (modern) constellations are carefully examined on a celestial globe, and a line drawn round so as to encircle them, it is found that the boundary line is (roughly) a circle, and that the centre of this circle falls in the constellation of Hydrus, near to the star Alpha. Now, at first sight, it would seem that this will not give us any information why this part of the heavens was not divided into constellations in ancient times. But when we find that the South Pole of the earth at one time pointed to the very centre of this circle containing the modern constellations, which was the case 4,000 years ago, it lets us see that, consequently, the stars composing these constellations then being near the South Pole, would in a particular latitude never appear above the horizon, and would thus be always invisible ; just as in Britain we never see the constellations near the South Pole at present, so that these stars could not have been seen by those who named the star-groups, and would thus remain unnamed till some future people would travel further to the South and see those stars, now the modern constellations.

As this circle is about 40 degrees in radius, all the stars that would be invisible to those who named the constellations would be within a distance of 40 degrees from the South Pole, so that if we find a latitude where the stars, situated within 40 degrees of the pole, are invisible, it will probably be the latitude of the country whose inhabitants first named the star-groups. This latitude would likely be 40 degrees North, but as the stars cannot be distinctly seen very near the horizon, we may safely say that the latitude would be a few degrees to the South of this, so as to make the Southern stars appear a little higher in the heavens. Thus the probable latitude where the star-groups were first named, would be about 36 or 37 degrees North, which is the latitude of Chaldæa ; and this, too, points to the Chaldæans as being the people who first divided the heavens into constellations.

We must, however, return to the description of some of the Southern constellations. Near Crux is a small group of stars called Musca, the bee ; and to the left of the cross is the constellation of Centaurus ; the two bright stars seen on the Milky Way both belong to this group, the one nearest the cross is β, and the other a, which is the nearest star to our solar system, as at present known. Below Centaurus is Lupus, the wolf ; while between Lupus and Octans will be seen the Southern Triangle. Underneath Triangulum are the constellations Pavo, the peacock, Ara, the altar, and part of Sorpio ; while to the right of Octans will be seen Hydrus, the water snake, and that part of the river Eridanus, which we never see, with its bright star *Achernar.* Directly above Hydrus are two small constellations, Dorado, the sword fish, and Reticulum, the net. Near the top of the Map will be seen the large and conspicuous constellation of Argo, which contains many bright stars, the principal one being *Canopus.* Near Dorado will be seen one of the Magellanic Clouds, this one is called Nebecula Major, the other one, seen near Hydrus, is called Nebecula Minor. When these nebulous masses are examined with the telescope, they are found to consist of small stars, clusters, and nebulæ of every description. The Milky Way is particularly bright in the Southern Hemisphere, and it will be seen that in this part of its course it has many branches, or off-shoots.

SHOOTING STARS.

WHEN one is looking at the heavens on a clear evening he is sometimes startled by a train of light suddenly coming into view, and gliding across the sky with great rapidity and then disappearing. This is a meteor or shooting star. They may be seen any clear night, and on certain dates,—notably the 13th and 27th of November, 10th August and the 20th April,—they are seen in very great numbers. These falling stars have from time immemorial attracted attention, and many theories have been advanced as to their real nature. At first it was generally supposed that shooting stars were nothing more than decomposed fluids floating in the higher strata of the atmosphere, which, after reaching a considerable height, became ignited, and thus appeared as falling stars. This accounts for the old belief that a display of shooting stars was a sign of stormy weather, it being further thought that as meteors were of atmospheric origin, so they would produce atmospheric disturbances, such as gales &c. Then it was supposed that shooting stars were the same as aerolites, or meteoric stones, and further that they had been ejected from volcanoes in the moon.

It is only, however, within the last fifty or sixty years, during which Meteoric Astronomy has been studied, that it has been found that the tracks of shooting stars seen on certain dates, when traced backwards, point to a particular part of the heavens, or that the meteors appear to radiate from a certain point (called the "radiant point") among the stars. Fig. 5 shows the position of the radiant point of the shooting stars seen on the nights of November 13th, 14th, and 15th. It will be noticed that these meteors appear to radiate from a point which is situated in the constellation of the Lion; on which account these shooting stars have been called *Leonides*. For a similar reason the meteors seen about the 10th of August, and 27th of November, are called *Perseids* and *Andromedes* respectively.

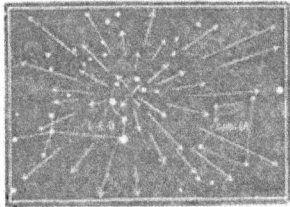

Fig. 5. The Radiant Point of the November Meteors.

This proved that meteors had no connection whatever with the earth or the moon, or, in other words, their celestial origin was demonstrated. The more reasonable hypothesis, which is now known to be accurate, was then advanced—viz., that meteors were small planetary bodies revolving in an orbit round the sun; but having a very eccentric path, the earth would be nearer at times to the ring of these bodies than at others, and in some cases would actually pass through it. Thus, when the earth was near to the ring of meteoric bodies, its great attractive influence would draw many of them down to it; and as they neared it their velocity would be momentarily accelerated, so that by the time they reached the atmosphere their speed would be so great that it would generate an immense heat—caused by the friction of the air—a friction which is generally more than sufficient to consume them before they can reach the surface of the earth. If any of these meteoric stones are large enough they will not be wholly burnt, but part of them will arrive at the earth's surface at an extremely high temperature. Many of these aerolites have been found in various parts of the earth, and some have been picked up shortly after they had fallen and while yet hot. In our museums at the present day there are seen specimens of meteoric stones, which, probably have travelled through space for many *millions* of years before they came within the attractive influence of our earth, and were thus brought down to us.

Lately the orbits of meteors have been discovered to be identical with the paths of known comets, and are thus found to be immense masses of stones, and meteoric dust following in the train of comets revolving round the sun. It is probable, too, that these comets with their trains of meteors, have at one time been ejected from other suns, or from the stars. For the meteors now belonging to our system would not likely be expelled from our sun, as any matter ejected from the sun would either be entirely carried away from our system, or if not thrown away from him with sufficient force, it would return to his glowing surface, so they must have travelled from other suns, till coming within the sphere of our sun's attraction, their paths would be so altered that they would henceforth revolve round him. These meteors, then, are actually parts of the stars. Now as the stars are situated at inconceivable distances from each other, the comets with their meteor trains, though sometimes travelling at great velocities, must have taken *millions* of years to journey from one star to the other; and as they probably would make many journeys from star to star before being caught by our sun, the number of years since they were expelled from a star may be so great that the sun, from which they were ejected, may now be cold and dead.

ON any dark and clear night, when there is no moon, a large band of cloudy light will be seen stretching right across the heavens from one side to the other. This is called the Milky Way. The position of this luminous zone in the sky, for any particular time, will be found by referring to the monthly Maps. In our latitudes it will be seen passing through, or near, the constellations, Scorpio, Sagittarius, Aquila, Cygnus, Cepheus, Cassiopeia, Perseus, Auriga, Gemini, and Canis Major. That part of its course extending from Canis Major to Auriga, is exceedingly faint and regular in outline; while, on the other hand, the part which traverses the constellations of Scorpio, Sagittarius, Aquila, and Cygnus, is more conspicuous, and in appearance very irregular. That portion of the Milky Way which is invisible to us, from being in the Southern hemisphere, passes through the constellations Argo, Crux, Ara, and Centaurus. In this part of the heavens, occupied by these constellations, it shines more conspicuously than it does in the Northern hemisphere. In the Ship, Argo, the brighter portion of the stream is completely divided by a dark and curiously-shaped opening, and near the Cross it is very bright, though some of the luminous patches, seen in the constellation of the Archer, may be even more brilliant.

When examined with a powerful telescope, the Milky Way, especially in those parts of its course where its appearance is very conspicuous, is truly a magnificent sight.—(Fig. 6). Even with so small an instrument as an opera glass, the vast number of stars that are rendered visible, is well calculated to impress the least thoughtful mind with a sense of the omnipotent power, and the infinite wisdom of the Creator, who brought so many *suns* into existence (for each of the many million stars seen in the Milky Way is a sun just like our own sun), so we do not wonder that, when the celebrated Schroeter was observing a part of the galaxy near Arided (α) in Cygnus, with a reflecting telescope nineteen inches in diameter, he was so impressed with the infinite grandeur of the scene, that it drew from him the natural exclamation, "What Omnipotence!"

But even before the telescope was invented, the Milky Way had attracted considerable attention; in fact, centuries before that instrument was even thought of, ancient astronomers had watched the luminous band as it stretched across the heavens, and had even arrived at the conclusion that its milky whiteness would, in all probability, be produced by the combined lustre of myriads of stars, which being situated at such vast distances away,

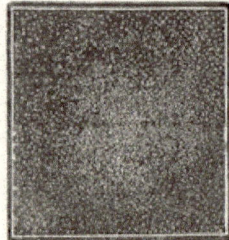

Fig. 6. Small part of the Milky Way as seen with a Telescope.

became individually indistinguishable; and thus Ovid says that "It is a road whose ground-work is of stars," and Manilius uses similar language.

It was only, however, when Galileo first turned the telescope to the Milky Way that its true nature became known. That instrument revealed to him the glorious assemblage of stars of all orders of brightness; from those which appeared to him as bright as the leading stars in the heavens, down to the smallest visible point of light, which could only be but momentarily glanced at as they glittered like a sprinkling of diamond dust against the background of the sky.

But there were some parts of the Galaxy that Galileo's telescope utterly failed to penetrate, and there still remained in the background that same misty light which had for so many centuries engaged the attention of astronomers; though the telescope that Galileo possessed must have revealed to him an enormous number of stars compared to what he had only before seen with the naked eye. With every increase of telescopic power more stars were seen, and greater depths were reached, but only to find, as Galileo had found, that still some parts would require a more powerful instrument to reveal the individual stars, that, by being so closely crowded together caused this cloudy light. And even when Sir William Herschel applied his powerful reflectors—the largest of which was 40 feet long and 4 feet in diameter—to this part of the heavens, and reached still farther depths, there was yet seen that same milky light which speaks of the myriads of stars still to be revealed. Nay, even Lord Rosse with his gigantic telescope—which is 6 feet in diameter and 54 feet long—could not resolve some of the luminous patches scattered throughout the Milky Way, though in some directions he penetrated vast distances into the infinite space, and saw stars so distant that the light from them must have taken many *hundreds* of years to journey to our earth, and travelling, too, at such a tremendous velocity that in a little over every five seconds it moves through a space equal to *one million* of miles. But though all the stars composing the Milky Way have not yet been revealed, still it has long been considered as demonstrated, that the milky light is caused by myriads of faint stars.

15

In the following Table the names of the principal Stars, that are visible in Britain, are given, corresponding to their letter in the Greek alphabet.—

Constellation.	Letter.	Name.
Andromeda	α	Alpheratz
	β	Mirach
	γ	Almach
Aquarius	α	Sadalmelik
	β	Sadalsund
	δ	Skat
Aquila	α	Altair
	β	Alshain
	γ	Tarazed
Aries	α	Hamal
	β	Sheratan
	γ	Mesartim
Auriga	α	Capella
	β	Menkalinan
Bootes	α	Arcturus
	β	Nekker
	ε	Izar
Canes Ven.	α	Cor Caroli
Canis Major	α	Sirius
	β	Mirzam
Canis Minor	α	Procyon
	β	Gomeisa
Capricornus	α	Secunda Giedi
	δ	Deneb Algiedi
Cassiopeia	α	Schedar
	β	Chaph
Cepheus	α	Alderamin
	β	Alphirk
	γ	Errai
Cetus	α	Menkar
	β	Diphda
	ζ	Baten Kaitos

Constellation.	Letter.	Name.
Cetus	ο	Mira
Columba	α	Phact
Corona Bor.	α	Alphecca
Corvus	α	Alchiba
	δ	Algores
Crater	α	Alkes
Cygnus	α	Arided
	β	Albireo
Draco	α	Thuban
	β	Alwaid
	γ	Etanin
Eridanus	β	Cursa
Gemini	α	Castor
	β	Pollux
	γ	Alhena
	δ	Wasat
	ε	Mebsuta
Hercules	α	Ras Algethi
	β	Korneforos
Hydra	α	Alphard
Leo	α	Regulus
	β	Denebola
	γ	Algeiba
	δ	Zosma
Libra	α	Zuben el Genubi
	β	Zuben el Chamali
	γ	Zuben Hakrabi
Lyra	α	Vega
	β	Sheliak
	γ	Sulaphat
Ophiuchus	α	Ras Alhague
	β	Cebalrai

Constellation.	Letter.	Name.
Orion	α	Betelgeux
	β	Rigel
	γ	Bellatrix
	δ	Mintaka
	ε	Alnilam
Pegasus	α	Markab
	β	Scheat
	γ	Algenib
	ε	Enif
	ζ	Homan
Perseus	α	Mirfak
	β	Algol
Pisces	α	Kaitain
Piscis Aust.	α	Fomalhaut
Sagittarius	α	Kaus Australis
Scorpio	α	Antares
Serpens	α	Unukalhai
Taurus	α	Aldebaran
	β	Nath
	η	Alcyone (Pleiad)
Ursa Major	α	Dubhe
	β	Merak
	γ	Phecda
	δ	Alioth
	ζ	Mizar
	η	Benetnasch
		Talitha
Ursa Minor	α	Polaris
	β	Kochab
Virgo	α	Spica
	β	Zavijava
	ε	Vindemiatrix

THE GREEK ALPHABET.

Letter.	Name.	Letter.	Name.	Letter.	Name.	Letter.	Name.
α	Alpha	η	Eta	ν	Nu	τ	Tau
β	Beta	θ	Theta	ξ	Xi	υ	Upsilon
γ	Gamma	ι	Iota	ο	Omicron	φ	Phi
δ	Delta	κ	Kappa	π	Pi	χ	Chi
ε	Epsilon	λ	Lambda	ρ	Rho	ψ	Psi
ζ	Zeta	μ	Mu	σ	Sigma	ω	Omega

MAP I.

THE CONSTELLATIONS FOR OCTOBER AND NOVEMBER.

NORTH

EAST

WEST

SOUTH

The above Map is a representation of the Starry Heavens in the Evening at the following Dates and Hours :—

OCTOBER	16 at	...	...	10.20	NOVEMBER	6 at	...	9.0	NOVEMBER	26 at	...	...	7.40
	21 ,,	...	...	10.0		11 ,,	...	8.40	DECEMBER	1 ,,	...	...	7.20
	26 ,,	...	...	9.40		16 ,,	...	8.20		6 ,,	...	...	7.0
	31 ,,	...	...	9.20		21 ,,	...	8.0		11 ,,	...	...	6.40

The circular boundary of this Map represents the Horizon, with the principal points of the compass indicated. The Cross in the centre is the zenith or that part of the sky which is directly overhead.
A list of Arabic names of the Principal Stars corresponding to the Greek letters in each Constellation is given on page 16.

DESCRIPTION OF MAP I.

I N this Map, the Great Bear, Ursa Major, will be seen due North and at the lowest part of its diurnal course; it is easily recognised by its seven bright stars, (Fig. 7) four of which form a somewhat elongated square. [This is considered to be the most splendid and conspicuous of those constellations in our latitudes which never set. It is now seen to its best advantage. The Egyptians called this constellation the Hippopotamus (the bear being unknown in Egypt), but it was recognised as a bear by the Greeks, Persians, &c., and when America was discovered, the Indians in the Northern part of that country knew this constellation as the (Polar) Bear, showing that they had independently recognised it themselves. It has, however, many other names besides the Great Bear, such as David's Car, Charles' Wain, (or waggon), &c; but the most popular name by which it is known in this country is the Plough. This constellation was used long before the invention of the Mariner's Compass to guide the paths of ships at night, as Manilius informs us—

> "Seven equal stars adorn the Greater Bear
> And teach the Grecian sailors how to steer."

The two bright stars forming the right side of the square are called the 'Pointers' (see Fig. 7), because a line passing through these stars in an upward direction points out a bright star in the tail of the Little Bear, named *Polaris*, the Pole Star.] Hanging down below Polaris and to the left is the constellation of Ursa Minor; this group is somewhat the same shape as the Great Bear, but it appears in an inverted manner. Midway between the two Bears will be seen the long constellation of Draco, the dragon, one of the oldest asterisms; γ, β, and ξ represent the head of the dragon, γ and β forming the eyes. Cepheus is above the Pole Star; the three stars α, β, and γ forming a curved line. A line from *Alioth* (ϵ) in the Great Bear to Polaris, and continued about the same distance on the other side will point out the conspicuous constellation of Cassiopeia, which is nearly overhead; this group is easily identified by its five brightest stars forming the letter W. Gemini, the twins, is in the North-East rising; this constellation is easily distinguished by two bright stars (α and β) one above the other, the top star is called *Castor*, the bottom one *Pollux*. Above Castor is Auriga, the waggoner, its brightest star, *Capella*, is now midway between the horizon and the zenith. Nearly due East will be seen Orion, three conspicuous stars in a line, Orion's belt, are just above the horizon. Immediately above Orion

Fig. 7. The Plough

is Taurus, the bull, with its bright ruddy star *Aldebaran*, and above Aldebaran is a conspicuous cluster of stars called the 'Pleiades' belonging to the same constellation. Perseus is midway between Taurus and Cassiopeia, and on the Milky Way, which is very bright in this region of the heavens. In the South-East is the long constellation of Cetus, the whale; below Cetus is the river Eridanus, while above it will be seen the first constellation of the Zodiac—Aries, the ram.

Looking now due South, a large square—formed by four bright stars—will be seen between the point overhead and the horizon; this is called the 'Square of Pegasus,' though only three stars belong to that constellation. [The star forming the North-East corner of the square was formerly called both δ Pegasi and α Andromedæ, but astronomers only retain the name of α Andromeda, from which circumstance there is now no star called δ of Pegasus. This constellation of the flying horse (Pegasus) was anciently called 'Nimrod's Steed;' its principal stars are *Markab* (α), *Scheat* (β), and *Algenib* (γ).] Andromeda, the chained lady, will be seen above and a little to the left of the square of Pegasus, while Triangulum will be seen between Andromeda and Aries; and between Cetus and Pegasus will be seen the barren constellation of Pisces, the fishes. Below the square of Pegasus, and very low down on the horizon is the bright star *Fomalhaut*, the principal star in Pisces Australis, the southern fish.

In the South-West and a little above the horizon is Aquarius, the water-bearer, with Capricornus, the goat, underneath; these are both constellations of the Zodiac. Nearly due West and at a low elevation will be seen three bright stars in a line, the brighter star between the other two is *Altair*, the principal star in Aquila, the Eagle. Above Aquila is Cygnus, the swan; this constellation is a very conspicuous one, and is easily found out by its brightest stars forming a huge cross. Directly above Altair there is a small group of stars called the 'Dolphin' or Delphinus, and to the left of this group there is another small constellation, Equleus, the horse.

A very conspicuous star will be seen to the North of West, this is the star *Vega* in Lyra, the harp, the brightest star in the Northern heavens. Part of Ophiuchus, the serpent-bearer, is directly below Lyra. Hercules and Bootes are in the North-West low down on the horizon; and between Bootes and Hercules will be seen the star *Alphecca* in the Northern Crown.

17

DESCRIPTION OF MAP II.

THE Great Bear has now moved a little to the East of the North, the Pointers being uppermost. Car Caroli is just above the Northern horizon, shining brightly. The Little Bear hangs down below the Pole Star; while Draco twists half way round the Pole, its tail passing between the two Bears, and its head under the right foot of Hercules. Virgil, *vide* Dryden, says:—

> "Around our poles the spiry dragon glides,
> And like a wand'ring stream, the Bears divides."

[The Alpha (*Thuban*) of this constellation was nearest to the North Pole of the heavens upwards of 4,000 years ago; it was then only 10' distant from the polar point,—equal to one-third the apparent diameter of the moon,—a point that will not be approached by our present Pole Star nearer than about three times that distance, equal to the apparent diameter of the moon.* The Gamma of Draco (*Etanin*) is a star that passes very near the zenith of Greenwich when due South; from which circumstance Bradley, the Astronomer-Royal, in 1725 thought that this star would be favourably situated for determining its annual parallax (*i.e.*, the angle that the earth's orbit would make, as seen from the star), and so find the distance from the star to the earth. In this he was disappointed, as the angle was too minute to be easily detected, but he made the wonderful discovery of the aberration of light, which is the best proof that the earth is travelling in an orbit round the sun.]

Turning to the North-East, the head of Leo, the lion, will be seen rising, and a long way to the right will be seen Canis Minor, the little dog, its principal star *Procyon* being just above the Eastern horizon. The feet of the Twins, Gemini, are immediately above Procyon—Castor being directly above Pollux—and above Gemini is Auriga, with its conspicuous star Capella shining brightly.

The whole of the constellation of Orion is now fairly above the horizon, his 'Star-gemmed Belt' shines very conspicuously. Taurus is above Orion, with Aldebaran and the Pleiades, which are now seen very plainly. Above Taurus, and nearly overhead, is Perseus on the Milky Way. [There is a remarkable variable star in this constellation called *Algol* (*i.e.*, the Demon); this star undergoes a considerable change in its brightness, for in about 4 hours it gradually diminishes in lustre, from between a second and third magnitude, until it appears as a fourth magnitude star; it remains as such for about 18 minutes, and then in the next 4 hours recovers its brightness in a like gradual manner, and retains it for the remaining part of its period—viz., 2 days, 12½ hours. The period in which all these variations are performed is 2 days, 20 hours, 49 minutes.]

In the South-East, and a little above the horizon, is the river Eridanus, one of the old asterisms; the most conspicuous star in this constellation, *Achernar*, is never seen in Britain, as it is far to the South, so that it is always below our horizon; but the next important star, *Cursa* (β), is easily found from its close proximity to Rigel, in Orion, being just above that star. Cetus is now due South, and in its best position for being well seen. Directly above Cetus are the constellations Pisces, Aries, and Triangulum. Andromeda and Pegasus are in the South-West, the great square of Pegasus being somewhat tilted up; below it will be seen Aquarius just about to dip below the horizon.

A little to the North of the zenith will be found the W of Cassiopeia; this constellation is supposed to have a resemblance to a seated lady—the wife of King Cepheus. [In November 1572 there appeared suddenly an exceedingly bright star near κ of Cassiopeia—its exact position will be seen in Fig 8. While Tycho Brahe, the celebrated Danish astronomer, was returning from his observatory to his house, he saw a number of people gazing at a very bright star, which he was certain had not been visible an hour before. It surpassed all the stars in brilliancy, and even the planet Jupiter when brightest; it remained visible for 16 months, and then gradually disappeared. As other stars were seen near the same place in the years 945 and 1264, it is supposed that these were appearances of the same star, so we may not unreasonably expect the return at any time of this remarkable visitant. For, from 945 to 1264 there are 319 years, and from 1264 to 1572 there are 308 years, so that if we take the average interval,

Fig 8.—Cassiopeia, with the new star.

313 years, and add this to 1572 it gives us the year 1885, as about the time when another visitation might be expected.]

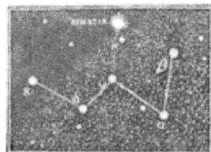

Underneath Cassiopeia will be seen the cross of Cygnus, standing upright; while to the right of it and nearer the horizon is Lyra with its bright star, Vega. Cepheus is above the head of Draco; below the dragon's head is Hercules, which is now very low down on the North-Western horizon.

* The accurate distance of Polaris from the Celestial Pole, when at its nearest to it, is 26 minutes, 30 seconds, which will take place in the year 2095, A.D.

MAP II.

THE CONSTELLATIONS FOR NOVEMBER AND DECEMBER.

The above Map is a representation of the Starry Heavens in the Evening at the following Dates and Hours :—

NOVEMBER 16 at	...	...	10.20	DECEMBER 6 at	...	...	9.0	DECEMBER 26 at	...	...	7.40
21 ,,	...	...	10.0	11 ,,	...	...	8.40	31 ,,	...	...	7.20
26 ,,	...	...	9.40	16 ,,	...	...	8.20	JANUARY 5 ,,	...	...	7.0
DECEMBER 1 ,,	...	...	9.20	21 ,,	...	...	8.0	10 ,,	...	...	6.40

The circular boundary of this Map represents the Horizon, with the principal points of the compass indicated. The Cross in the centre is the zenith or that part of the sky which is directly overhead.
A list of Arabic names of the Principal Stars corresponding to the Greek letters in each Constellation is given on page 16.

MAP III.

THE CONSTELLATIONS FOR DECEMBER AND JANUARY.

The above Map is a representation of the Starry Heavens in the Evening at the following Dates and Hours :—

DECEMBER 16 at	...	...	10.20	JANUARY 5 at	...	...	9.0	JANUARY 25 at	...	...	7.40
21 „	...	...	10.0	10 „	...	...	8.40	30 „	...	...	7.20
26 „	...	...	9.40	15 „	...	...	8.20	FEBRUARY 5 „	...	...	7.0
31 „	...	...	9.20	20 „	...	...	8.0	10 „	...	...	6.40

The circular boundary of this Map represents the Horizon, with the principal points of the compass indicated   The Cross in the centre is the zenith or that part of the sky which is directly overhead.
A list of Arabic names of the Principal Stars corresponding to the Greek letters in each Constellation is given on page 16.

DESCRIPTION OF MAP III.

THE dragon, Draco, is now due North ; Bootes and Hercules being directly underneath, and partly below the horizon. The Great Bear has moved to the North-East, the tail hanging down. The Little Bear is still below the Pole Star, and above Draco. Between the North-East and the East Leo will be seen nearly above the horizon, with its brilliant star *Regulus*, the lion's heart. Observe the sickle group in this constellation, Regulus being at the handle. In the East, and exactly between the horizon and the zenith, will be seen the two conspicuous stars belonging to Gemini ; and above these, and nearly overhead, is Auriga with its bright Capella. Procyon, in Canis Minor, is below the Twins, between the East and the South-East ; while between Procyon and Leo is the small constellation of Cancer, the crab. Canis Major, the great dog, is rising in the South-East, the principal star *Sirius* will be seen sparkling with great brilliancy. Orion is nearly standing upright between the South and the South-East ; below it is the small constellation of Lepus, the hare. To the right and above Orion is Taurus, the bull, the second constellation of the Zodiac ; *Aldebaran* represents the eye of the bull, the stars β and ζ the two horns. The beautiful cluster of stars called the *Pleiades* forms part of this constellation ; six stars are distinctly seen, though more are visible with good eyesight—Fig. 9 is a view of this cluster as seen with a small telescope. This celebrated group of stars is mentioned by Job, and there is another instance of its being noticed 3,400 years ago. Near Aldebaran another group will be noticed, but not so conspicuous as the Pleiades, this is called the Hyades.

Eridanus is now South, and nearly South-West is Cetus, the whale, the most extensive group in the heavens. The star in the neck of the whale is called *Mira*, the epithet 'wonderful,' being given to it on account of its great variation in brilliancy. It shines as a second magnitude star, and decreases in lustre till it becomes invisible to the naked eye in a period of about 300 days. Above Cetus is Aries the first constellation of the Zodiac. Manilius thus describes it :—

Fig. 9.—The Pleiades.

> " First Aries, glorious in his golden wool,
> Looks back and wonders at the mighty bull."

[More than 2,000 years ago the middle of this constellation was on one of the Equinoctial Points (*i.e.*, the points where the Equinoctial cuts the Ecliptic), from which it has been called 'the first point of Aries ;' but owing to the precession of the Equinoxes, or the motion of the axis of the earth round a perpendicular to the Ecliptic, the Equinoctial Points have moved to the right on the Ecliptic about 25,900 years.* Thus Aries 2,000 years ago occupied the same position with regard to the Equinoctial as Pisces now does, but since then it has *apparently* moved to the left a whole sign ; so that the *constellation* of Aries is now in the *sign* of Taurus, Taurus in Gemini, &c. *Mesertim*, the γ of Aries is slightly interesting from the fact that it was one of the first double stars that was discovered. Dr. Hook while observing a comet in 1664 accidently came across this star with his telescope and found that it consisted of two stars close together. There are now several *thousands* of double stars, not only detected, but their apparent distances accurately measured.]

Above Aries is Triangulum, and nearly overhead is Perseus. In the West will be seen Pegasus just about to sink below the horizon, the diagonal of the square formed by a Andromedæ, and a Pegasi being vertical. Between Pegasus and Cetus, Pisces will be seen ; while above the square of Pegasus is Andromeda, the chained lady. Cassiopeia has now moved away from overhead towards the North-West ; and underneath Cassiopeia are the constellations of Cygnus and Lyra, both being just above the horizon, while Cepheus is midway between Vega and the Zenith.

* The Ecliptic is divided into 12 equal parts which are called the 'signs of the Zodiac ;' they were named more than 2,000 years ago, after the constellations that occupied the same part of the Ecliptic as the signs ; but since then, the signs have moved away from the constellations. The signs are named as follows :—

| ♈ Aries. | ♊ Gemini. | ♌ Leo. | ♎ Libra. | ♐ Sagittarius. | ♒ Aquarius. |
| ♉ Taurus. | ♋ Cancer. | ♍ Virgo. | ♏ Scorpio. | ♑ Capricornus. | ♓ Pisces. |

DESCRIPTION OF MAP IV.

N the North, the bright star Vega will be seen tipping the horizon. The Dragon's head is now exactly North, and above Hercules. Ursa Minor is still below Polaris, but it is now beginning to assume a horizontal position. The Great Bear is in the North-East, between the point overhead and the horizon, the tail hangs down pointing to Bootes, which is rising. To the right of the Great Bear's Tail is Canis Venatici, the greyhounds; the only bright star in this constellation is called *Cor Caroli*. The whole of Leo has now risen above the horizon, and Virgo is just beginning to make its appearance; while to the right of Leo and nearly South-East will be seen the winding constellation of Hydra, with its leading brilliant *Alphard*, the 'solitary one.' Above Alphard is Cancer, and to the right of Cancer is Procyon, in the little dog. Above Procyon is Gemini, which is the third constellation of the Zodiac—it was originally represented as a pair of kids, but the Greeks altered them into two children, named *Castor* and *Pollux*. Below the feet of the Twins, and near the horizon, will be seen the blazing Dog-star, *Sirius*, which is the brightest of all the stars. [Sirius must be a sun of an immense size, as it has been found to be situated at a distance at least 20 times further away than the nearest star, and yet it shines so brilliantly—in fact, it is thought that to equal it in size it would require no less than 2000 suns the same as ours. This star has been found to be travelling through space away from us, at the enormous velocity of 20 miles per *second*. From very minute irregularities in the annual motion of Sirius it was long thought that there would probably be a dark body revolving round it, and this remarkable idea has been found to be correct, for a small companion has been discovered by means of one of the powerful American telescopes, it being only seen with great difficulty, as it probably shines by reflected light.]

To the left of Sirius, part of the Ship, Argo, will be seen; and to the right of Sirius is Lepus. Above Lepus is Orion, which is now a little past the South and at his greatest elevation—the best position to be well observed. This is the most beautiful and brilliant of all the constellations, and in ancient times there was none more noted than Orion, the 'mighty giant.' Manilius, the astronomical poet of the ancients, thus writes about Orion:—

> "Orion's beams! Orion's beams!
> His star-gemmed belt, and shining blade;
> His isles of light, his silvery streams,
> And gloomy gulfs of mystic shade.

The three stars in a line, near the centre of the group, are called the 'Belt,' and by Job the 'Bands of Orion;' it is known, however, by many other names, such as the 'Three Kings,' the 'Ell-and-Yard,' 'Jacob's Staff,' &c. The star α in Orion is called *Betelgeux*, it is a very irregular variable star, and forms with Procyon and Sirius a large equilateral triangle. [Directly underneath the middle star in the belt, there will be seen a star surrounded with a haze, this is the great Nebula of Orion, one of the most wonderful objects in the heavens. A very small telescope will suffice to show it as a luminous cloud with small stars shining through it; but it requires one of the huge telescopes that are now employed by astronomers to see it to advantage. When seen with one of these instruments, it is truly a magnificent and wonderful sight; the whole field of view is filled with an irregular mass of green shining mist, which is apparently broken up into flocculent masses, delicate clouds of light, sprays and wisps, and standing out from the cloudy background, like a sprinkling of diamond dust, are seen faint glittering stars. The real nature of this mysterious object long remained unknown, for telescopes without number have been turned to it in the hope of resolving its misty light into stars, but each have failed in turn. Even the 40-feet reflector of Sir W. Herschel, which was four feet in diameter, could not reveal the individual stars that were supposed to produce this luminous glow. Nay, even the gigantic telescope of Lord Rosse, with its six-feet mirror, could not resolve it completely, but nevertheless, it was thought that with a slightly more powerful instrument, it would be seen as a galaxy of stars. Astronomers, therefore, were rather taken by surprise when Mr. Huggins announced that, by means of the spectroscope, it was demonstrated that this object was nothing more than a great mass of incandescent gas, such as hydrogen, nitrogen, &c. Since this wonderful discovery many astronomers now believe that it consists of the material from which suns and systems of worlds will, at some future time, be made.]

To the right of Orion is Eridanus, and a little further in the same direction is Cetus, now dipping below the horizon. Taurus is above Orion; while above Taurus and nearly overhead is the bright Capella in Auriga. Aries has now reached the West; a little to the right of Aries is Andromeda, the stars α, β, and γ being vertical, while below Andromeda, and partly underneath the horizon is Pegasus. Perseus has now moved from overhead towards the West; while Cassiopeia and Cepheus are in the North-West, about midway between the zenith and the horizon—Cepheus being directly above the cross of Cygnus, which has partly set.

MAP IV.

THE CONSTELLATIONS FOR JANUARY AND FEBRUARY

The above Map is a representation of the Starry Heavens in the Evening at the following Dates and Hours :—

JANUARY	15 at	...	...	10.20	FEBRUARY	5 at	...	...	9.0	FEBRUARY	25 at	...	...	7.40
	20 „	...	...	10.0		10 „	...	...	8.40	MARCH	2 „	...	...	7.20
	25 „	...	...	9.40		15 „	...	...	8.20		7 „	...	...	7.0
	30 „	...	...	9.20		20 „	...	...	8.0		12 „	...	...	6.40

The circular boundary of this Map represents the Horizon, with the principal points of the compass indicated. The Cross in the centre is the zenith or that part of the sky which is directly overhead.
A list of Arabic names of the Principal Stars corresponding to the Greek letters in each Constellation is given on page 16.

MAP V.

THE CONSTELLATIONS FOR FEBRUARY AND MARCH

NORTH

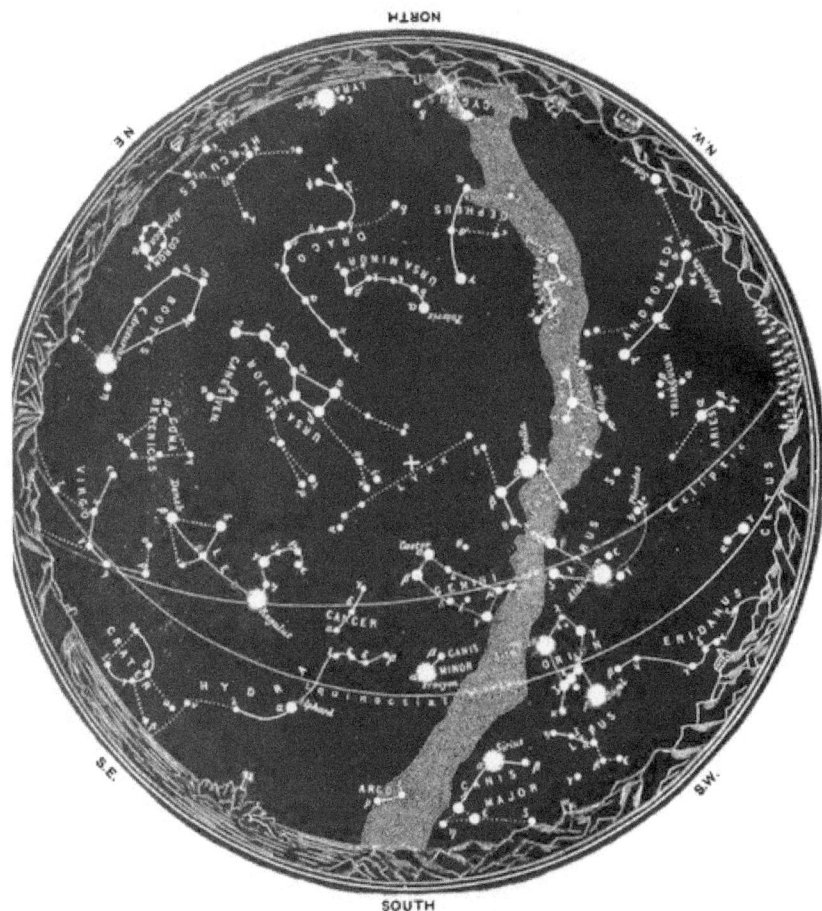

SOUTH

The above Map is a representation of the Starry Heavens in the Evening at the following Dates and Hours :—

FEBRUARY	15 at	...	...	10.20	MARCH	7 at	...	...	9.0	MARCH	27 at	...	...	7.40
	20 „	...	...	10.0		12 „	...	...	8.40	APRIL	1 „	...	...	7.20
	25 „	...	...	9.40		17 „	...	...	8.20		6 „	...	...	7.0
MARCH	2 „	...	...	9.20		22 „	...	...	8.0		11 „	...	...	6.40

The circular boundary of this Map represents the Horizon, with the principal points of the compass indicated. The Cross in the centre is the zenith or that part of the sky which is directly overhead.
A list of Arabic names of the Principal Stars corresponding to the Greek letters in each Constellation is given on page 16.

CYGNUS is now exactly North, only part of the cross being visible; to the right of it will be seen the sparkling Vega, just above the horizon, while still further in the same direction is Hercules, with Draco and the Little Bear directly above — the latter being now almost horizontal. The stars β and γ in this constellation are sometimes called the 'Guardians of the Pole.' Ursa Minor is one of the most important constellations in the heavens, because its leading star, *Polaris*, is situated so near the North Pole, there is no other star of more value to the sailor, as from it he can not only find the cardinal points, but, by measuring its apparent height above the horizon he can find his distance from the Equator, or his latitude, as it is called. In ancient times when there was no compass to guide the seaman, it was of still greater use, in fact it was invaluable, as without it no vessel could have safely sailed out from the sight of land. Thus Dryden describes the infancy of navigation :—

> " Rude as their ships were navigated then,
> No useful compass or meridian known,
> Coasting, they kept the land within their ken,
> And knew no North but when the Pole-star shone. "

The Great Bear is rapidly approaching the zenith; Corona Borealis, the northern crown, and Bootes, the herdsman, being directly underneath, between the North-East and the Eastern horizon. Coma Berenices is to the right of Arcturus and nearly East—this constellation contains many clusters of small stars and nebulæ; it was formerly part of Leo. The whole of Virgo has not yet risen above the horizon, that part of it which is visible will be seen due East. The tail of Leo is directly above Virgo, the whole of this constellation being now nearly South-East, with its leading brilliant Regulus. The Sickle group, formed by the stars α, η, γ, ζ, μ, and ϵ, is now in an upright position.

Crater, the cup, is directly underneath Leo, and very low down on the horizon, the star *Alkes* (α) being exactly below Regulus and above β. Hydra is to the right of the cup; it will be seen winding upwards through Alphard and nearly to the Little Dog. Above the head of Hydra is Cancer, the crab, the fourth Zodiacal constellation. It contains no conspicuous stars, but it has a noted star-cluster which is called the Bee-hive—the Præsepe of the ancients. This small group is very easily found, for a line from Pollux—the lower of the two brilliant stars in the Twins—to Regulus will, about midway between these stars, pass very near to it. When this cluster of stars, is seen with a telescope, it is a magnificent sight, each sparkling like so many diamond points against the sky. Fig. 10 is a view of this interesting cluster as seen with a telescope.

Gemini will be seen a little to the West of South; its two principal stars, Castor and Pollux, now being on the meridian (i.e., the line that runs exactly from North to South). Underneath Pollux is the bright star Procyon, and to the right of it and above will be seen *Gomeisa* the β of Canis Minor. [Procyon, like Sirius, is thought to have one of those dark bodies revolving round it, of which there are probably many thousands in the universe, and, of course, being invisible, they can only be detected by their attractive influence on other stars. The regular variations in the brightness of the star *Algol* (in Perseus) are thought to be caused by one of these dark bodies revolving round it, as when between us and

Fig. 10.—The Præsepe, or Bee-hive.

that star it will intercept some of the light, so the star will shine with diminished lustre. These invisible bodies are probably nothing else than extinct suns, which at one time would be as brilliant and as active as our own, but as their energy has long ceased, so they are now cold and dead, and such it is thought will be the case with our own sun in some 17 millions of years hence.]

Between the South and the South-West will be seen Canis Major, with its brilliant star Sirius—the star to the right of it is called *Mirzam*. To the right of the Dog-star is Lepus, and above Lepus is the conspicuous Orion, now nearing the South-West. Part of Eridanus has disappeared, and nearly the whole of Cetus—only the head of the whale being now visible. Above Eridanus is the ruddy Aldebaran, with the comet-like Pleiades to the right of it; while above Aldebaran is the brilliant Capella, the leading star in Auriga, the waggoner; the bright star to the left of Capella is called Menkalinan. Between Auriga and the North-Western horizon are the constellations of Perseus and Andromeda—the bright stars α, β, and γ in the latter constellation being now nearly vertical. The whole of Pegasus has sunk below the horizon, with the exception of one star, *Scheat*. To the left of Andromeda will be seen Triangulum and Aries. The W of Cassiopeia has now assumed this position—Σ; it will be seen to the right of Perseus, and in the North-West, while Cepheus is to the right of Cassiopeia, and below the Pole Star.

DESCRIPTION OF MAP VI.

EPHEUS will be seen due North, *Errai* (γ) being a little below the Pole Star. Ursa Minor now twists a little upwards towards the right; while directly underneath it, and partly below the horizon, is Cygnus. Lyra is to the right of Cygnus, and in the North-East rising. Draco is directly above Vega, the tail circling upwards towards the zenith. Between Vega and the East is Hercules, and still further to the right is part of Serpens; while nearly East and above Serpens is Corona Borealis, the northern crown. [It will be noticed that this constellation has a striking resemblance to a crown; it is thought that originally it belonged to Bootes, though undoubtedly an old constellation itself; the brightest star in it is called *Alphecca*, which will be seen sparkling like a gem in this 'celestial coronet.' To the left of Alphecca, and in a position where no star was visible to the naked eye, there appeared very suddenly in th: year 1866 a new star shining with great brilliancy, and which in due time rapidly disappeared again; the true position of this star will be seen in Fig. 11. It was believed among astronomers that there had taken place on this new star (or *sun*) some tremendous disturbance, caused, probably, by the downfall of some mighty mass of matter, which had the effect of making it shine out with about nine *hundred* times its former brilliancy! But may we not anxiously enquire what would likely be the result if a conflagration like that which undoubtedly took place on this remote sun, were at any time to happen to our sun? Would any of the life on earth ever hope to survive such an awful change in heat and light as this? Not only would all the various forms of life on earth be utterly destroyed, but on all the other members of our solar system there would be such a change effected, that if any life existed, even on the remote Neptune * (which is not thought to be the case at *present*), it would at once be completely extinguished. Nay, even if the heat and light of our sun were to increase, not nine hundred, but only nine times its present amount, all the higher orders of life on the earth would be completely destroyed; and if any of the various forms of life were to survive, these would only be some of the lower orders existing in caves, with a few of the lowest forms of vegetation. And thus it is thought (with good reason, too), that probably the life that existed on the whole system of worlds that circled round this distant star must have been annihilated, and as the heat and light of this star increased so very suddenly, there could have been given but short warning to the inhabitants of these worlds].

Fig. 11.—The Northern Crown, showing the New Star.

Above the Northern Crown is the noted constellation of Bootes, the herdsman, with its leading brilliant Arcturus, which is considered to be one of the brightest stars in the heavens.

The grey-hounds, Canes Venatici, will be seen above Arcturus; while directly overhead is Ursa Major. In the South-East is the brilliant *Spica*, now fairly above the horizon; the whole of Virgo will now be seen, and also Corvus, which is to the right of Spica: above Corvus and a little to the right is Crater, and directly above this group is Leo, the lion, the fifth constellation of the Zodiac, its brightest star *Regulus*, the lion's heart, being just above the Ecliptic, while between it and Arcturus will be seen Denebola (β), the lion's tail. Underneath Regulus and towards the right is the solitary star Alphard, in Hydra, the sea serpent.

The greater part of Canis Major has set, but its brightest star Sirius is still seen, it being now South-West, just above the horizon. Above Sirius is the Little Dog with Procyon; while to the right of Procyon will be seen the feet of Gemini, the twins, which constellation is now in an upright position. Below the Twins will at once be seen the noble Orion, now somewhat tilted over, and just about to sink below the Western horizon; the three stars in the belt are lying in a horizontal position; while below them is the bright Rigel, and above, the ruddy Betelgeux. To the right of the latter will be noticed another red star, which is Aldebaran, in Taurus, the bull; in fact from this star appearing so very red, and lying near the path of the planets, it has often been mistaken by inexperienced observers for the ruddy planet Mars. To the right of Aldebaran, and at nearly the same elevation, will be seen the Pleiades; while above this cluster is Auriga with its conspicuous Capella, to the right of which is Perseus; while below this latter and to the right is Andromeda, part of this constellation having now disappeared; and between the North-West and the North, and a little above the horizon, will be seen the W of Cassiopeia.

* Neptune is the outermost known planet in the solar system, and it is situated over 30 times further away from the sun than the earth, so that the heat and light it receives from the sun will be nearly 1,000 times less than what we receive.

MAP VI.

THE CONSTELLATIONS FOR MARCH AND APRIL.

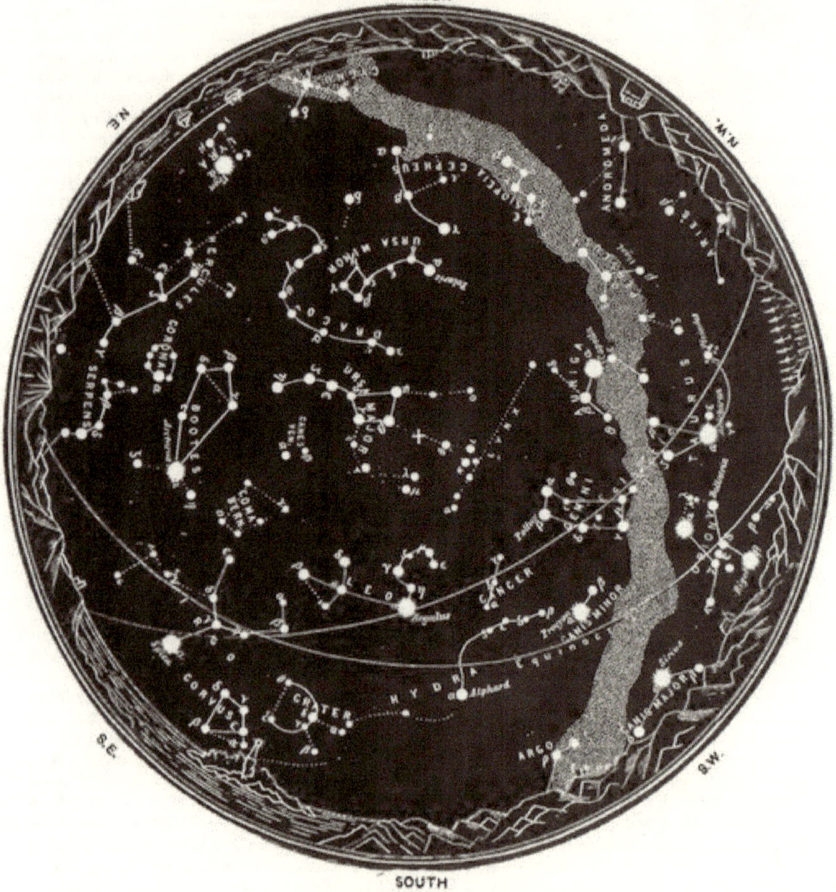

The above Map is a representation of the Starry Heavens in the Evening at the following Dates and Hours:—

MARCH	7	at	...	...	11.0	MARCH	27	at	...	...	9.40	APRIL	16	at	...	...	8.20
	12	„	...	...	10.40	APRIL	1	„	...	...	9.20		21	„	...	...	8.0
	17	„	...	...	10.20		6	„	...	...	9.0		26	„	...	...	7.40
	22	„	...	...	10.0		11	„	...	...	8.40	MAY	2	„	...	...	7.20

The circular boundary of this Map represents the Horizon, with the principal points of the compass indicated. The Cross in the centre is the zenith or that part of the sky which is directly overhead.
A list of Arabic names of the Principal Stars corresponding to the Greek letters in each Constellation is given on page 16.

MAP VII.

THE CONSTELLATIONS FOR APRIL AND MAY.

The above Map is a representation of the Starry Heavens in the Evening at the following Dates and Hours :—

APRIL	6 at	...	...	11.0	APRIL	21 at	...	...	10.0	MAY	7 at	...	...	9.0
	11 ,,	...	...	10.40		26 ,,	...	...	9.40		12 ,,	...	...	8.40
	16 ,,	...	...	10.20	MAY	2 ,,	...	...	9.20		17 ,,	...	...	8.20

The circular boundary of this Map represents the Horizon, with the principal points of the compass indicated. The Cross in the centre is the zenith or that part of the sky which is directly overhead.
A list of Arabic names of the Principal Stars corresponding to the Greek letters in each Constellation is given on page 16.

DESCRIPTION OF MAP VII.

CASSIOPEIA is now nearly North and between Polaris and the horizon, while Cepheus has already moved to the right of the North, γ being exactly below the Pole Star. The Little Bear is above Cepheus, the guardians of the Pole, β and γ, being above Polaris and to the right. Cygnus is exactly North-East, the great cross lies horizontal. Directly above Cygnus is Draco; while above the foot of the Cross of Cygnus will be seen the bright sparkling Vega, to the right of which is the old constellation of Hercules, which is supposed to represent a man kneeling. Its leading star α, which is not a very conspicuous one, will be seen due East, it is called *Rasalgeti*, from the Arabian Rás-al-gàthi, the kneeler's head. A little below this star is the α of Ophiuchus, which is named Rasalhague, from an Arabian word meaning the serpent-bearer's head. This constellation of Ophiuchus, the serpent-bearer, is one of the old asterisms, only part of it is at present visible. Above the latter is Serpens, the serpent, and above the serpent's head, formed by the stars β and γ, is Corona Borealis; while to the right of the Northern Crown will be seen the brilliant Arcturus, the leading star in Bootes. Directly underneath Arcturus, and above the South-Eastern horizon is Libra, the balance; while underneath it part of Scorpio is beginning to make its appearance. To the right of Libra and nearly South is Virgo, the sixth constellation of the Zodiac; it contains a very conspicuous star called *Spica*, which will be seen between the South and the South-East.

Corvus, the crow, and Crater, the cup, are now South, and near the horizon; these are both old constellations: δ of the former is called *Algorib*, and α of the latter *Alkes*. Above Spica, is Coma Berenices, and Canes Venatici; Cor Caroli in the latter being exactly between Arcturus and the zenith. The Great Bear is still overhead, and, in passing, we may notice how useful this constellation is for pointing out many of the leading stars which are now visible. The stars α and β point out Polaris; δ and γ point out Regulus, in the lion; δ and α point out Capella, in Auriga; δ and β point out Castor and Pollux; a line from Regulus to η in the tail of the Bear, and continued about the same distance on the other side, points out Vega; and a line through the stars ζ and η passes very near Arcturus.

Leo has now passed the South, Deneb in the tail being exactly on the meridian; above Regulus will be seen the star λ, which is mentioned because close to this star there is situated the radiant point of the famous 13th of November meteors, or the point in the heavens from which these meteors appear to burst, and on which account these shooting stars are familiarly known as the *Leonids*.

Alphard in Hydra, is now South-West; while to the right of it the bright star Procyon, in the little dog, will be seen. Cancer is above the head of Hydra, while to the right of it is Gemini, now rapidly nearing the horizon. Orion has all but disappeared, Betelgeux and Bellatrix alone are now visible. To the right of Orion is Taurus, the ruddy Aldebaran in the bull's eye is just about to sink below the horizon; and nearly North-West is the Pleiades which will also soon disappear. This beautiful cluster of stars, which are so conspicuous when high up in the heavens, will now scarcely be noticed, from being so low down on the horizon. They are about to leave us for a season, and after having disappeared, they will not again be seen in the evening till the beginning of September, when they will reappear in the North-East about 10 p.m. They will, however, be seen in the early morning rising before the sun in about two months; so that they will be invisible for a short time, owing to the sun being in that part of the heavens, and hiding them by the overpowering brilliancy of his rays. Thus, Hesiod (who lived nearly 3,000 years ago) says, referring to the invisibility of the Pleiades :—

> "There is a time when forty days they lie,
> And forty nights, concealed from human eye,
> But in the course of the revolving year,
> When the swain sharps the scythe, again appear."

Above this interesting group is Capella and Menkalinan, both belonging to Auriga; while to the right of the latter is Perseus, and still further in the same direction and nearly North, is that part of Andromeda which is at present visible.

23

DESCRIPTION OF MAP VIII.

ASSIOPEIA is still in the North, Andromeda being underneath it; Cepheus is above and to the right of Cassiopeia; the stars α, β, and γ are now lying horizontal; the Little Bear is above the star γ of Cepheus; the Guardians of the Pole being above Polaris and slightly to the right.

In the North-East the whole of the Swan will be seen fairly above the horizon; the huge cross in this constellation is exactly horizontal, *Arided* being at the top and *Alberio* at the bottom. Observe the marked brilliancy of the Milky Way in this part of its course, and in contrast to it, notice the dark space between the two branches, from γ to β, which has been called the Northern 'Coal Sack.' Above Alberio, in Cygnus, is the small but conspicuous constellation of Lyra, the harp, with its bright sparkling Vega—this star is supposed to be a sun of enormous bulk, and situated at a vast distance away from us, it being thought that the light from it requires no less than 18 years to reach our earth.

Nearly East and low down on the horizon is Aquila, the eagle, its leading brilliant *Altair* will at once be noticed, as it lies between two smaller stars. To the left of Altair will be seen the small group of the Dolphin, which has just risen. Above Lyra and to the right is the kneeling Hercules, while underneath it is the upright Ophiuchus, triumphant over the serpent, Serpens, which constellation will be seen passing through the Serpent-bearer, the head of the serpent being on the right and below Corona Borealis. To the right of the Northern Crown we see Bootes, with its brilliant Arcturus. [This star is travelling through space at an almost inconceivable speed, in fact its apparent motion is so great that since the beginning of our era, it has moved over a space equal to $2\frac{1}{2}$ times the apparent diameter of the moon. Owing to this star having such a great motion it was long supposed that it was one of the nearest stars to our earth; but now it has been found that it is situated at a distance not less than seven times that of the nearest known star. According to Mr. Huggins it is travelling towards us at the enormous velocity of over 50 miles per *second*. The heating effect, too, of this star has been measured, it being found by Mr Stone that we receive as much heat from Arcturus as would be given out by a 3 inch cube of boiling water placed at a distance of 400 yards.]

Between the South and the South-East are two constellations of the Zodiac, Libra, the balance, and Scorpio, the scorpion— in the latter will be noticed the fiery red Antares (α). To the right of Libra is Virgo, its bright star Spica, being now past the South; while above Virgo is Coma Berenices. Corvus and Crater are now between the South and the South-West, low down on the horizon; while below these constellations, and twisting to the West will be seen Hydra, just above the horizon. At one time (more than 4,000 years ago) this constellation of the Sea Serpent coincided with the equinoctial, but, owing to the swaying motion of the earth's axis, the equinoctial has since moved away.

Leo is between the West and the South-West; above the tail of the Lion is Cor Caroli, in Canes Venatici, while nearly overhead and towards the West will be seen the Great Bear, *Benetnasch* (η), in the tail, being close to the zenith. To the right of Regulus, in the Lion, and due West is Cancer, the crab; while to the right of Cancer, the Twins will be seen setting.

In the North-West and a little above the horizon is the bright Capella, in Auriga; and to the right of it, Perseus will be seen, its interesting star Algol now being just above the horizon.

MAP VIII.

THE CONSTELLATIONS FOR MAY AND JUNE.

The above Map is a representation of the Starry Heavens in the Evening at the following Dates and Hours :—

MAY				MAY				JUNE			
2 at	...	...	11.20	17 at	...	...	10.20	1 at	...	...	9.20
7 ,,	...	...	11.0	22 ,,	...	...	10.0	6 ,,	...	...	9.0
12 ,,	...	...	10.40	27 ,,	...	...	9.40	11 ,,	...	...	8.40

The circular boundary of this Map represents the Horizon, with the principal points of the compass indicated.   The Cross in the centre is the zenith or that part of the sky which is directly overhead.
A list of Arabic names of the Principal Stars corresponding to the Greek letters in each Constellation is given on page 16.

MAP IX.

THE CONSTELLATIONS FOR JUNE AND JULY.

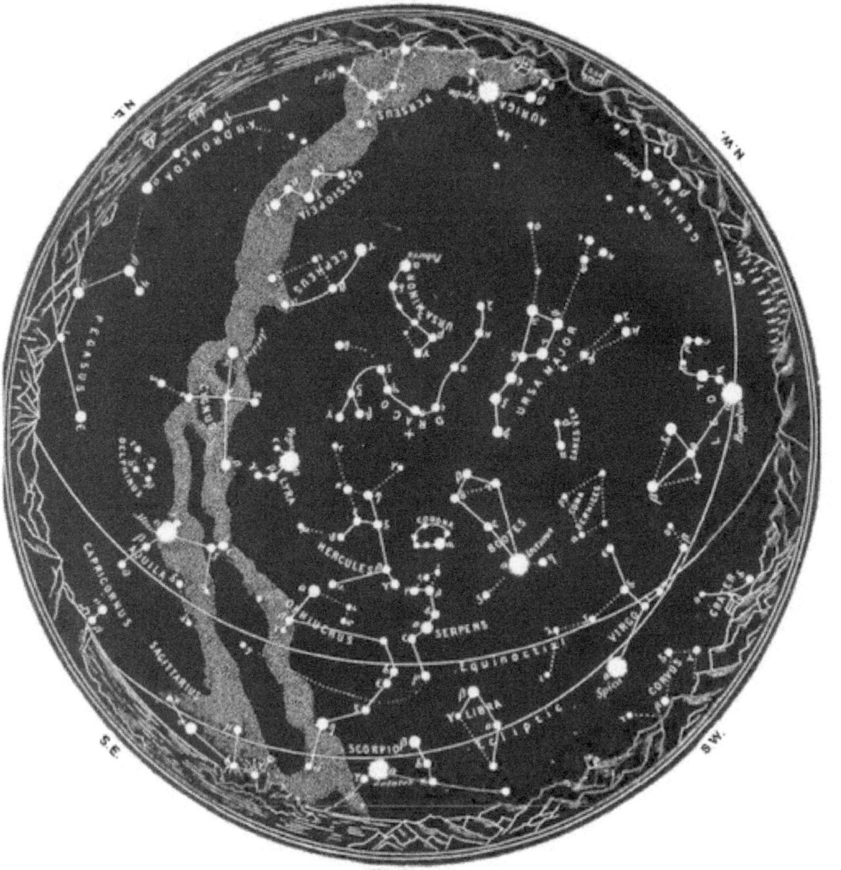

The above Map is a representation of the Starry Heavens in the Evening at the following Dates and Hours :—

JUNE				JUNE				JULY		
1 at	..	..	11.20	16 at	...	...	10.20	2 at	...	9.20
6 „	...	...	11.0	21 „	...	...	10.0	7 „	...	9.0
11 „	...	...	10.40	26 „	...	...	9.40	12 „	...	8.40

The circular boundary of this Map represents the Horizon, with the principal points of the compass indicated. The Cross in the centre is the zenith or that part of the sky which is directly overhead.
A list of Arabic names of the Principal Stars corresponding to the Greek letters in each Constellation is given on page 16

DESCRIPTION OF MAP IX.

IN the North we now see the constellations of Auriga and Perseus; while above Perseus and a little to the right is the conspicious **W** of Cassiopeia; directly above which is Cepheus; and underneath Cassiopeia and just above the horizon is Andromeda, the chained lady—in Grecian mythology these last four constellations are connected with each other, for Cepheus was a king of Æthiopia and Cassiopeia his queen. Andromeda was their daughter, who king Cepheus was obliged, in order to preserve his kingdom, to chain to a rock to be devoured by a sea-monster (which is represented by the constellation of Cetus); but Perseus, on his return from the conquest of the Gorgons, rescued her, and turned the monster into a rock by showing it the head of Medusa.

The Square of Pegasus is now beginning to make its appearance, that part of it which is now visible will be seen in the North-East and to the right of Andromeda. Cygnus is now due East, and midway between the horizon and the zenith; the cross is still horizontal; above Alberio (β) is the brilliant star Vega, in Lyra; while below it is the Dolphin, to the right of which is Altair, the principal star in the Eagle, Aquila. The star (γ) above Altair is called *Trazed*, the one below (β) *Alshain*. Underneath Aquila we see Capricornus beginning to make its appearance, the stars α and β alone being visible.*

In the South-East, part of Sagittarius, the archer, has come into view; while above it is Ophiuchus with Serpens, and above Ophiuchus is Hercules, reaching nearly overhead. Below the Serpent-bearer is Scorpio with the fiery Antares; while to the right of it is Libra, and still further in the same direction is Virgo, which will be easily distinguished by means of its brilliant Spica, now seen exactly South-West and a little above the horizon. Corvus and Crater are below Virgo rapidly setting; while directly above Spica is Arcturus in Bootes, and to the right of this star is Coma Berenices.

Leo is now West and rapidly approaching the horizon; above the tail of the Lion is Canes Venatici. The Great Bear has now fallen away from overhead towards the North-West, the tail being uppermost: exactly North-West and just about to disappear are the twin stars, Castor and Pollux, all that is now visible of Gemini.

To the right of Ursa Major and nearly overhead is Draco; while underneath it and twisting upwards above Polaris will be seen the Little Bear.

* Owing to the strong twilight in the month of July, and as the map represents the heavens at an hour which will not be dark enough for observing the less conspicuous stars, so there may be inconvenience in finding some of the stars near the horizon; but if Map X. be used instead of this one, and still using the dates belonging to Map IX.—only two hours will require to be added to them —then the sun will be further below the horizon, and, of course, the stars will appear more brilliant.

DESCRIPTION OF MAP X.

AURIGA is still near the North, the bright star Capella will be seen scintillating above the horizon. Perseus has risen a little higher in the sky, and is now nearly North-East; while above it is Cassiopeia and Cepheus. To the left of Cepheus is Ursa Minor, which is still above the Pole Star, but it has now fallen towards the left.

Andromeda lies horizontal between the North-East and the East; while below it will be seen Triangulum. Pegasus is to the right of Andromeda; it is now due East and completely above the horizon; it will be noticed that the Great Square is tilted up, the diagonal formed by the stars Algenib and Scheat is nearly vertical.

Pisces is beginning to appear again, part of it being above the horizon and directly underneath Pegasus. To the right of Pisces is Aquarius, the water-bearer, the second last constellation of the Zodiac; while to the right of Aquarius, and a little beyond the South-East, will be seen another Zodiacal constellation—Capricornus, the goat. Above Capricornus is Aquila, with its leading brilliant Altair; to the left of which will be seen Delphinus, and above the Dolphin is the conspicuous Cross of Cygnus—Arided (a) being at the top, and Albireo (β) at the bottom. To the right of the Swan is Lyra; the bright Vega is now due South and high up in the heavens overhead; observe the conspicuous triangle formed by the stars Vega, Altair, and Arided.

In the South we see Sagittarius, part of it only being above the horizon; to the right of it and nearly South-West is Scorpio with its ruddy Antares, now nearing the horizon. Above Scorpio is Ophiuchus, to the right of which is Serpens; while above the Serpent is the Northern Crown. Hercules is above the Serpent-bearer, and above Hercules is the head of Draco, now exactly overhead; the tail of the Dragon will be observed twisting down towards the North.

Virgo will be seen setting in the West, the brilliant star Spica is rapidly sinking below the horizon. Coma Berenices is in the West above Vindemiatrix (ε of Virgo); while to the left of Coma Berenices, is Arcturus in Bootes. Part of Leo has now disappeared below the horizon, the remaining part of it which is visible will soon sink out of sight; above the star δ in the Lion will be seen Cor Caroli, to the right of which is the Great Bear, now North-West and midway between the horizon and the zenith.

MAP X.

THE CONSTELLATIONS FOR JULY AND AUGUST

The above Map is a representation of the Starry Heavens in the Evening at the following Dates and Hours :—

JULY				JULY				AUGUST			
JULY	7 at		11.0	JULY	22 at	...	10.0	AUGUST	6 at	... ...	9.0
	12 „		10.40		27 „	...	9.40		11 „	... ...	8.40
	17 „		10.20	AUGUST	1 „	...	9.20		16 „	... ...	8.20

The circular boundary of this Map represents the Horizon, with the principal points of the compass indicated. The Cross in the centre is the zenith or that part of the sky which is directly overhead.
A list of Arabic names of the Principal Stars corresponding to the Greek letters in each Constellation is given on page 16.

MAP XI.

THE CONSTELLATIONS FOR AUGUST AND SEPTEMBER.

NORTH

SOUTH

The above Map is a representation of the Starry Heavens in the Evening at the following Dates and Hours :—

AUGUST	6 at	...	...	11.0	AUGUST	26 at	...	...	9.40	SEPTEMBER	16 at	...	...	8.20
	11 ,,	...	...	10.40		31 ,,	...	...	9.20		21 ,,	...	...	8.0
	16 ,,	...	...	10.20	SEPTEMBER	6 ,,	...	...	9.0		26 ,,	...	...	7.40
	21 ,,	...	...	10.0		11 ,,	...	...	8.40	OCTOBER	1 ,,	...	...	7.20

The circular boundary of this Map represents the Horizon, with the principal points of the compass indicated. The Cross in the centre is the zenith or that part of the sky which is directly overhead. A list of Arabic names of the Principal Stars corresponding to the Greek letters in each Constellation is given on page 16.

DESCRIPTION OF MAP XI.

HE bright Capella, in Auriga, is now nearly North-East, and well above the horizon; to the right of it we see Perseus; below which will be noticed the Pleiades, low down on the horizon, which shows us that Taurus is about to make its appearance for the winter months. Above Perseus is Cassiopeia, and above, and to the left of Cassiopeia, is Cepheus, now nearly overhead; the branching of the Milky Way near α (*Alderamin*) is well worthy of notice.

Aries is to the right of the Pleiades; while above it is Triangulum and Andromeda. Pegasus, which has attained a fair elevation, is now between the East and the South-East; the Great Square is still tilted up, Scheat being nearly above Algenib; below the Square Pisces will be seen—this constellation, which is not a very conspicuous one, consists of two fishes linked together by a ribbon, the one fish lies parallel with the bottom side of the Square of Pegasus, and the other with the left side of the Square. Below *Enif*, in Pegasus, is Aquarius, now at a fair height above the horizon, while to the right of the Water-Bearer is Capricornus, now nearly South.

Above Capricornus we see the conspicuous Altair, in the Eagle; to the left of which is Delphinus. Above the Dolphin, and nearly overhead is the brilliant Cross of Cygnus, now exactly South.

[Some years ago astronomers were startled by the sudden appearance of a conspicuous star to the left of Arided in the Swan, in a place where no star had been visible to the naked eye before. This strange star was first noticed by Professor Schmidt of Athens, on the 24th of November 1876: it obtained the same brightness as a star of the third magnitude, and then gradually faded away, and disappeared, till it required a telescope to reveal it. The exact position where this star appeared will be seen in Fig. 12.

Fig. 12.—Cygnus, with New Star.

In the same Fig. there will be seen a small star marked 61 (a double star); this is the noted 61 Cygni—the nearest known star to us in the Northern hemisphere. With regard to it Webb, in his "Celestial Objects," says:—" These suns were the first of the host of heaven to reveal to Bessel, in 1838, the secret of their distance. This is, probably, 366,400 times that of the earth from the sun—itself 92,400,000 miles—a space so vast that the light, which reaches us from the sun in 8 minutes, employs nearly 6 years to traverse it.'. . . How vast must be the dimensions of this great Universe! What a temple for the Creator's glory! 'All the whole heavens are the LORD's'—those heavens are crowded with millions of millions of stars; and of all that countless multitude, millions, probably, for one, are at a distance incalculably exceeding that of 61 Cygni!"]

Between the South and the South-East Sagittarius will be seen setting; and to the right of it is Ophiuchus rapidly approaching the horizon. Serpens is to the right of Ophiuchus; while above the head of the Serpent is Hercules. The brilliant Vega, in Lyra, has now passed the South, and is gradually sinking lower in the heavens.

To the right of Hercules is Corona Borealis, and further in the same direction is Bootes, its leading star Arcturus being to the right of the West, and near the horizon.

In the North-West we see Cor Caroli, and above it the Great Bear, which is now nearing the North, but not like other constellations to disappear below the horizon, for, as Homer mentions,* it is "denied to slake its beams in Ocean's briny baths," or, in other words, to set, or sink below the horizon. Above Ursa Major is Draco, which has now fallen away from overhead; to the right of Draco will be seen the Little Bear; the Guardians of the Pole being to the right of Polaris.

---

* In his description of the shield of Achilles, on which there was engraved—

"The Heaven, the Sea,
The Sun that rests not, and the moon full-orb'd,
There, also, all the stars which round about
As with a radiant frontlet bind the skies,
The Pleiades and Hyades, and the might
Of the huge Orion, with him Ursa call'd,
Known also by his popular name, the Wain,
Which wheeling round the Pole still looks toward
Orion; only star of these denied
To slake his beams in Ocean's briny baths."

DESCRIPTION OF MAP XII.

URSA MAJOR is near the North; that part of this constellation which is called the Plough now being between the North and the North-West, and at a fair elevation above the horizon. Below the tail of the Bear is Canes Venatici; above it is Draco, and above the Dragon is Ursa Minor, which will now be seen to the left of the Pole Star; directly above Polaris and nearly overhead is Cepheus; and below Cepheus and the to right is Cassiopeia, the seated lady.

In the North-East we now see Auriga with its brilliant Capella well above the horizon; between Capella and Cassiopeia is Perseus, the rescuer of Andromeda, with the head of Medusa. Between the East and the North-East Taurus will be seen, with its leading star, the ruddy Aldebaran, and its conspicuous cluster of stars, the Pleiades; the whole of this constellation of the Bull has now risen clearly above the horizon, which reminds us that the winter constellations are again coming round.

Stretching along the horizon from the East to the South-East we see the Whale, Cetus, with its remarkable variable star Mira. In Grecian mythology this constellation represents the great sea-monster that Neptune sent to devour Andromeda, but Perseus interfered in time to save her. Above the head of Cetus is Aries with Triangulum; while above the Triangle is Andromeda; below which, and to the right of Aries, is Pisces.

Nearly South we see the winged horse Pegasus; the Great Square being in the South-East, and midway between the horizon and the zenith. Directly beneath the neck of the Flying Horse and just tipping the horizon will be seen Fomalhaut, the brightest star in Pisces Australis; above it we see Aquarius; to the right of the Water-Bearer, Capricornus will be noticed, and exactly South-West is Sagittarius, which has all but disappeared.

In the South-West and at a high elevation will be seen the conspicuous Cross of Cygnus, now standing up-right; below it and to the left is Altair in Aquila. To the right of the Cross of Cygnus is the brilliant Vega in Lyra, which is now due West and exactly between the point overhead and the horizon; below Vega is Hercules, and underneath Hercules is Ophiuchus with Serpens. Between the West and North-West is Corona Borealis; below and to the right of which is Arcturus, in Bootes, just above the horizon; while above Bootes is Draco, now exactly North-West.

MAP XII.

THE CONSTELLATIONS FOR SEPTEMBER AND OCTOBER.

The above Map is a representation of the Starry Heavens in the Evening at the following Dates and Hours :—

SEPTEMBER 11	at	...	...	10.40	OCTOBER	1	at	...	...	9.20	OCTOBER	21	at	...	...	8.0
26	,,	...	...	10.20		6	,,	...	...	9.0		26	,,	...	...	7.40
21	,,	...	...	10.0		11	,,	...	...	8.40		31	,,	...	...	7.20
26	,,	...	...	9.40		16	,,	...	...	8.20	NOVEMBER	6	,,	...	...	7.0

The circular boundary of this Map represents the Horizon, with the principal points of the compass indicated.   The Cross in the centre is the zenith or that part of the sky which is directly overhead.
A list of Arabic names of the Principal Stars corresponding to the Greek letters in each Constellation is given on page 16.